High-Temperature Electrochemical Energy Conversion and Storage

Fundamentals and Applications

CRC Press
Taylor & Francis Group
6000 Broken Sound Parkway NW, Suite 300
Boca Raton, FL 33487-2742

First issued in paperback 2019

ISBN-13: 978-1-4987-7927-2 (hbk)
ISBN-13: 978-0-367-88983-8 (pbk)

Library of Congress Cataloging-in-Publication Data

Names: Shi, Yixiang, author.
Title: High-temperature electrochemical energy conversion and storage : fundamentals and applications / Yixiang Shi [and 3 others].
Description: Boca Raton : CRC Press, Taylor & Francis Group, 2018. | Series: Electrochemical energy store & conversion | Includes bibliographical references.
Identifiers: LCCN 2017038439| ISBN 9781498779272 (hardback : acid-free paper) | ISBN 9780203701935 (ebook)
Subjects: LCSH: Direct energy conversion. | High temperature electrolysis. | Energy storage.
Classification: LCC TK2896 .S55 2018 | DDC 621.31/242--dc23
LC record available at https://lccn.loc.gov/2017038439

Visit the Taylor & Francis Web site at
http://www.taylorandfrancis.com

and the CRC Press Web site at
http://www.crcpress.com

Contents

Foreword

High-temperature solid oxide fuel cells (SOFCs) are electrochemical devices for converting hydrocarbon fuels into electricity, and have attracted great interest in the past few decades for clean and efficient distributed power generation. I began researching this technology in the early 1970s at the Westinghouse Electric Corporation in Pittsburgh, Pennsylvania. There, as manager of Fuel Cell Technology from 1984 to 2000, I was responsible for the development of SOFCs for stationary power generation. In this role, I led an internationally recognized group in SOFC technology and brought this technology from a few-watt laboratory curiosity to fully-integrated 200 kW size power generation systems. In the year 2000, I joined the Pacific Northwest National Laboratory, a U.S. Department of Energy laboratory managed by Battelle, in Richland, Washington, as a Battelle Fellow and Director, Fuel Cells; there I provided senior technical, managerial, and commercialization leadership to the laboratory's extensive fuel cell and other clean energy programs, and helped launch the U.S. Department of Energy's new national program on SOFCs called the Solid State Energy Conversion Alliance (SECA). I got to know Professor Yixiang Shi, one of the coauthors of this book, about 10 years ago when I was a distinguished visiting professor under China's 1000 Talent Program at the China University of Mining and Technology-Beijing. Because of my more than four decades of experience with SOFCs and the promise that Professor Shi has shown in this technology, I am delighted to write this foreword for his book *High-Temperature Electrochemical Energy Conversion and Storage: Fundamentals and Applications*.

In the last two decades, great effort has been devoted to improve the reliability, robustness, and endurance of SOFCs, lower the capital cost of SOFC systems, and develop SOFC systems for residential, transportation, and military sectors; these efforts have been funded by several national programs in the United States, Japan, Europe, and China. Noteworthy among these are the programs funded by the U.S. Department of Energy in the United States, the New Energy and Industrial Technology Development Organization (NEDO) programs in Japan, the Fuel Cells and Hydrogen Joint Undertaking (FCH JU) programs in the European Union, and National Key Research and Development Programs by the Ministry of Science and Technology (MOST) in China. In addition, industry has also funded SOFC research, development, and demonstration programs.

The reverse operation of SOFC, namely, solid oxide electrolysis cell (SOEC), is another important high temperature electrochemical technology. With the help of external power sources, SOECs can split CO_2 and H_2O into CO and H_2, which can be converted to useful fuels and other chemicals. The operation of SOEC, or reversible SOFC systems, enables grid scale energy storage and facilitates emergence of smart power grids. On the other hand, turning CO_2 into fuels and chemicals with renewable power is a promising way to reduce overall carbon emissions.

SOFC and SOEC systems have reached a point where systematic design of the products is more difficult than cell fabrication technology in which great strides have been made through efforts via various national research programs. To achieve proper system configuration, deep understanding of reaction processes and accurate calculation of energy balances of the reaction system are required. This book provides comprehensive, up-to-date knowledge on theoretical principles, experimental phenomena, reaction mechanisms, and simulation methods on SOFCs and SOECs. The authors begin this book by discussing the fundamentals and applications of high-temperature electrochemical energy conversion and storage. Chapter 1 provides a brief introduction to a variety of high-temperature electrochemical energy conversion and storage technologies, including solid oxide fuel cell, molten carbonate fuel cell, solid oxide electrolysis cell, molten metal battery, and solid oxide metal-air redox battery. Chapters 2 and 3 discuss solid oxide fuel cells and solid oxide electrolysis cells, including their working principles, modeling, and systems. Chapters 4 and 5 focus on two kinds of specific SOFCs: flame fuel cells and solid oxide direct carbon fuel cells; both theoretical and experimental aspects are discussed based on the experiences the authors gained through their recent research. Readers will find this book helpful as it provides a comprehensive and clear picture of the achievements in high-temperature fuel cells and batteries that should help in furthering the understanding of underlying high temperature electrochemistry. The book will also broaden knowledge on fuel cell/battery systems and ignite further interest in this field.

Since my first foray in this technology in the early 1970s, high temperature solid oxide technology has advanced at a fast pace for energy conversion and storage. This book is timely, and the information presented is significant and fundamental enough for it to be useful as a textbook for graduate and senior undergraduate students interested in this technology as well as benefit scientists, engineers, and other researchers already active in the field.

Subhash C. Singhal, PhD, MBA
Battelle Fellow and Director, Fuel Cells (Emeritus), Pacific Northwest National Laboratory, USA Academician, U.S. National Academy of Engineering Founding Chair, Biennial International Symposia on Solid Oxide Fuel Cells *Co-Editor of the highly acclaimed book,* High Temperature Solid Oxide Fuel Cells: Fundamentals, Design and Applications, *2003 Richland, Washington*

Preface

Electrochemical energy conversion and storage devices are considered a promising power technology that can directly convert chemical energy in fuel into power. Especially, high-temperature electrochemical devices have numerous existing or potential applications, including fuel cells for power generation, production of high-purity hydrogen or syngas by electrolysis, high-purity oxygen by membrane separation, as well as several different types of high-temperature batteries, e.g., molten metal battery, high-temperature metal–air battery, high-temperature flow battery, etc.

In this book, we discuss the fundamentals and applications of high-temperature electrochemical energy conversion and storage, which are the authors' main research contents. In Chapter 1, we provide a brief introduction to a variety of high-temperature electrochemical energy conversion and storage technologies, including solid oxide fuel cell (SOFC), molten carbonate fuel cell, solid oxide electrolysis cell, molten metal battery, and solid oxide metal–air redox battery. Chapters 2 and 3 discuss SOFCs and solid oxide electrolysis cells, including their working principles, experiments, modeling, and systems. Chapters 4 and 5 focus on two specific kinds of SOFCs: flame fuel cells and solid oxide direct carbon fuel cells. Experimental results and theoretical evaluation of both fuel cells based on the authors' own research are presented to the readers.

This book provides a comprehensive and clear picture of the achievements in high-temperature fuel cells and batteries. We also hope it would help our readers furthering the understanding of high temperature electrochemistry. We sincerely hope this book will also widen knowledge on fuel cell/battery systems and ignite more interest on the topic.

Acknowledgments

The authors would like to acknowledge the earnest contributions from the Group for Clean Energy Conversion and Utilization, Department of Thermal Engineering, Tsinghua University (Chapter 2: Tianyu Cao, Chapter 3: Yu Luo, Chapter 4: Hongyu Zeng, Yuqing Wang, Chapter 5: Tianyu Cao), and the efforts of Peidong Song in editing this book.

The authors would also like to acknowledge the support from Project 2014CB249201, which is supported by the National Basic Research Program of China (973 Program), Projects 51476092, 51576112 (National Natural Science Foundation of China, NSFC), and Youth Foundation Program for Fundamental Scientific Research in Tsinghua University (221 Program).

About the Authors

Dr. Yixiang Shi is an associate professor in the Department of Thermal Engineering at Tsinghua University, Beijing, China. Dr. Shi obtained his BA from the Department of Thermal Engineering, Tsinghua University, in 2003. He earned his PhD in 2008 from Tsinghua University and conducted related research in high-temperature fuel cells. During 2007–2008, he worked at the University of California, Irvine, as a joint educated PhD candidate. After that, he worked as a postdoctoral associate researcher in the Mechanical Engineering Department at MIT during 2008–2009. Dr. Shi currently serves as the deputy secretary in the Department of Thermal Engineering, Tsinghua University, and secretary of director of Key Laboratory for Thermal Science and Power Engineering of the Ministry of Education. He is engaged in the theoretical modeling and experimental characterization of the reaction and transport processes of solid oxide fuel cells, direct carbon fuel cells, direct flame fuel cells, elevated temperature pressure swing adsorption separation technology, and CO_2 electrochemical conversion. He is the author/coauthor of more than 100 peer-reviewed articles and several book chapters and patents. He was awarded the "221" Basic Research Plan for Young Faculties, Tsinghua University; Beijing Higher Education Young Elite Teacher Award; Outstanding award for Youth-Teaching Contest of Tsinghua University; Academic Excellence Young Faculties; Excellent Doctoral Dissertation; and Outstanding Ph.D. Graduation Award, as well as The academic youngster Scholarship of Tsinghua University. Dr. Shi is a board committee member of the International Academy of Electrochemical Energy Science (IAOEES).

Dr. Ningsheng Cai is a professor in the Department of Thermal Engineering at Tsinghua University, Beijing, China, and a distinguished professor of the Cheung Kong Scholars Programme of the Ministry of Education of China. Professor Cai now serves as the deputy director of the National Engineering Research Center of Clean Coal Combustion, Key Laboratory for Thermal Science and Power Engineering of the Ministry of Education, and Beijing Municipal Key Laboratory for CO_2 Utilization and Reduction. He is an editorial board member in Energy discipline for Springer. He earned his BS in power engineering from Xi'an Jiaotong University in 1982, ME in thermal energy engineering at Nanjing Institute of Technology in 1987, and PhDs in thermal energy engineering from Southeast University and in mechanical engineering at the University of Tennessee Space Institute (USA) in 1991. His current research activities cover several areas such as clean coal technologies, including CO_2 capture with solid adsorbents, chemical looping combustion, coal gasification combined cycle power generation and coproduction systems, solid oxide fuel cells, and electrolyte cells.

Tianyu Cao is a young researcher in electrochemical conversion and power generation. Tianyu received his bachelor degree from Department of Thermal Engineering at Tsinghua University in 2014. He then began to pursue a doctoral degree in the same department. Tianyu has been devoting himself to research and development of direct carbon fuel cell. He is also interested in reaction processes in fuel cells and rechargeable batteries.

Dr. Jiujun Zhang is a principal research officer and core-competency leader at the Energy, Mining and Environment, National Research Council of Canada (NRC-EME). He received his BS and MSc in electrochemistry from Peking University in 1982 and 1985, respectively, and his PhD in electrochemistry from Wuhan University in 1988. After completing his PhD, he took a position as an associate professor at Huazhong Normal University for two years. Starting in 1990, he carried out three terms of post-doctoral research at the California Institute of Technology, York University, and the University of British Columbia. Dr. Zhang has over 30 years of R&D experience in theoretical and applied electrochemistry, including over 18 years of fuel cell R&D (among these, 6 years at Ballard Power Systems and 12 years at NRC-IFCI (before 2011)/NRC-EME (after 2011)), and 3 years of experience in studying electrochemical sensors. Dr. Zhang holds several adjunct professorships, including one at the University of Waterloo, one at the University of British Columbia, and one at Peking University. Dr. Zhang has coauthored more than 400 publications including 230 refereed journal papers with approximately 16,000 citations, 15 edited/coauthored books, 11 conference proceeding papers, 36 book chapters, as well as 110 conference and invited oral presentations. He also holds more than 10 US/EU/WO/JP/CA patents and 11 U.S. patent publications and has produced in excess of 90 industrial technical reports. Dr. Zhang serves as the editor/editorial board member for several international journals as well as editor for the book series Electrochemical Energy Storage and Conversion (CRC Press). He is an active member of the Electrochemical Society (ECS), the International Society of Electrochemistry (ISE, fellow member), the American Chemical Society (ACS), the Canadian Institute of Chemistry (CIC), as well as the International Academy of Electrochemical Energy Science (IAOEES, board committee member).

1 Introduction to High-Temperature Electrochemical Energy Conversion and Storage

Due to the growing global demand for energy and the tightening regulation of carbon emissions, advanced power systems and energy storage devices are urgently needed. Electrochemical energy storage and conversion devices (abbreviated as EESC in this book) are the most promising power technologies for this purpose and can directly convert chemical energy stored in fuels or electroactive chemicals into electric power (an electroenergy conversion process), or convert electric power into fuels or electroactive chemicals (an electroenergy storage process). Generally, all batteries and supercapacitors can be classified as EESC devices because their charge operations are electroenergy storage processes, while discharging them can be regarded as consumption of energy stored. Having said this, however, low-/high-temperature fuel cells that normally convert fuels into electric power should be considered as electroenergy conversion devices, and electrolysis cells that normally convert electric power into fuels or chemicals should be considered as electroenergy storage devices.

The main advantages of EESC devices over other energy storage and conversion devices include (1) wide variety of applications in stationary, transportation, and portable microelectronics; (2) wide variety of power and energy density ranges; (3) mobile (wireless); (4) high storage-conversion efficiencies (40%–95%); (5) rechargeability; and (6) environmental friendliness. For example, the energy efficiency achieved in fuel cell systems can significantly surpass the Carnot limitation in combustion processes, which are commonly found in traditional combustion plants or generators. By using fuel cell systems, combustion-derived pollutants in direct combustion processes such as SO_x, NO_x, and soot can be eliminated.

The working processes of various fuel cells and batteries are similar to each other: the fuels supplied to the system or the electroactive chemicals stored inside the cell are oxidized at the anode in the anode chamber, while the oxidant is reduced at the cathode in the cathode chamber. The fuels (or electroactive chemicals) and the oxidant, instead of being mixed together, are separated with a gas-tight, ionic conducting membrane (serving as the electrolyte), which forces electrons released during fuel oxidation to power external circuits before it recombines with the oxidant. At the same time, ions are transported through the electrolyte to maintain the charge balance of the system.

EESC devices operated at high temperatures (ranging from 500°C to 900°C) require complex system configurations due to the need for thermal management.

High-temperature operations can however have benefits such as accelerated reaction kinetics, which can reduce or even eliminate the use of precious metal catalysts required for low-temperature EESC devices. For example, low-temperature fuel cells use expensive platinum (Pt)-based catalysts to activate electrode reactions, while high-temperature fuel cells, such as solid oxide fuel cells (SOFCs), can use relatively inexpensive metals such as nickel (Ni)-based or copper (Cu)-based catalysts. In addition, high operating temperatures can also improve the tolerance of fuel cells to impurities in the fuels. For example, in low-temperature fuel cells, such as proton exchange membrane fuel cells, even 10 ppm of CO or trace amount of NO_X/SO_X in the H_2 fuel or oxidant will cause a significant reduction in the lifetime of the Pt catalyst. This would increasing the cost of fuel/oxidant purification. In SOFCs, however, these impurities are not a problem, and in some cases, CO can even be used as a fuel. Furthermore, high-temperature electrochemical systems can use a wide variety of fuels. Currently, high-temperature fuel cells can operate on syngas, natural gas, and hydrocarbons; in fact, high-temperature fuel cells have even been fed with solid carbon.

High-temperature electrochemical energy devices have many applications, including electric power generation through fuel cells, high-purity hydrogen or syngas generation through electrolysis cells, high-purity oxygen generation through membrane separations, as well as electric power generation through several different types of high-temperature batteries, e.g., molten metal battery, metal–air battery, and flow battery.

High-temperature EESC devices face several challenges as well, including the stability of thermal management, high degradation rates of materials/components, as well as high costs. These challenges will be discussed separately in the following chapters of this book. The later chapters also provide recent progresses in high-temperature EESC development in terms of fundamental theories and industrial applications to the readers. People would benefit from the scientific and technical information presented.

1.1 INTRODUCTION TO SOLID OXIDE FUEL CELLS

SOFCs are typical high-temperature EESC devices. Most SOFC devices are employed as stationary generators for distributed power generation. Due to its elevated operation temperature, a wide variety of fuels can be fed into SOFC systems to generate power. For example, direct oxidation of hydrocarbon in SOFCs has been demonstrated by numerous researchers. Of the major components in an SOFC, such as the anode, the cathode, and the electrolyte, the electrolyte, made up with metal-oxide-based ceramics, is probably the most critical. In most cases, stabilized zirconia ceramic membranes (Yttrium Stabilized Zirconia (YSZ) or Scandium Stabilized Zirconia (ScSZ)) are ideal materials for the electrolyte in high-temperature SOFCs, as elemental doping into ZrO_2 lattices (Y or Sc) provides substantial oxygen ion conductivity above 800°C, and zirconia-based ceramics are stable under both reducing and oxidizing atmospheres.

SOFC stacks are often integrated with a steam reformer, which can convert natural gas (mostly methane) into smaller molecules such as H_2 and CO for fuel feeding in the fuel cells. The mechanisms of H_2 and CO oxidation in SOFC anodes need to be understood to improve the system efficiencies. In the literature, some elemental reaction mechanisms have been proposed and validated through both experimental results and numerical models. Coupling effects of the reactions and the transport processes have also been quantitatively evaluated. As observed, the electrochemical oxidation of CO at the SOFC anode is normally slower than that of H_2. Therefore, more efficient reformers are required to produce more hydrogen via hydrocarbon reforming and water–gas shift reaction, while the fraction of CO should be as small possible for better SOFC performances. In reforming reactors, partial oxidation reactions are initiated by a fuel-rich flame that at the same time serves as a heat source of the reforming reaction of CH_4. Direct coupling of the heat effects of different reactions can provide the reformer with higher reforming efficiencies and simpler configurations. An advanced reformer should have both high reforming efficiency and high system stability.

Fuel cell stacks, reformers, burners, and other balance of plant (BOP) devices are key components of an SOFC power system. Flue gas of SOFCs, often containing residue fuel species and substantial amounts of sensible heat, can be fully exploited through a bottom cycle. For smaller, household SOFC systems, cogeneration of heat and power (CHP) can be achieved. For a step further, tri-generation of heat, power, and cooling can be achieved if an adsorption chiller is integrated into the original fuel cell system. For example, larger, stationary combined-cycle-distributed power systems have been developed by Siemens–Westinghouse and Mitsubishi Heavy Industries (MHI). Both systems employ pressurized tubular SOFCs as the top cycle and microgas turbine as the bottom one. Exhaust gases from both the anode and the cathode chambers are ignited in the combustion chamber to drive the microturbine. Although tubular SOFCs possess lower power densities than planar SOFCs, it does show significant advantages due to its sealing feasibility and power scalability. Both tubular SOFC-based hybrid power systems can reach over 200 kW of power output (Siemens–Westinghouse: 220 kW; MHI: 250 kW). In the latest version of their SOFC hybrid system, the MHI system operating at 900°C can reach up to a power efficiency of 50.2% (lower heating value [LHV]) [1] with thousands of hours of operation while the whole system takes up an area of around 40 m^2 (3.2 × 3.2 × 12 m) [2]. This demonstrates that SOFC hybrid systems are a feasible technology for distributed power generation, as the 250 kW_{el} generated with heat production can fulfill the energy demands of a small community. This advanced power system with its low emissions, low noise, and acceptable power density can easily provide residents with clean power.

Although the SOFC systems mentioned earlier demonstrate feasibility, there are still technical challenges to overcome. One challenge is the controlling and integration of SOFC systems. Reformers, fuel cell stacks, bottom cycles, and other BOPs make SOFC a rigid system that requires careful design and operation. Carbon deposition at the anode is another major challenge during operation, as carbon growth on the anode during operation will result in cell delamination and loss of catalysts, leading to the failure of the whole cell. Therefore, understanding the mechanisms of carbon formation and consumption in SOFCs is of great importance. With respect to this, carbon-related reactions in SOFCs need to be understood to prevent carbon deposition on the anode.

As mentioned earlier, high operating temperature is the major cause of SOFC performance degradation. High temperatures result in rapid degradation of all components, and in particular, current collection and sealing. Therefore, most efforts in recent years have focused on lowering operating temperatures. Noticeable progresses have been made in selecting candidate materials for both intermediate-temperature SOFCs (IT-SOFC) and low-temperature SOFCs (LT-SOFC) [3,4]. Novel electrode and electrolyte materials that are functional under lower temperatures have also been developed. However, innovations in materials do not necessarily translate into improved technologies, and further developments and validation of practical SOFC power systems using these novel materials are still required.

1.2 INTRODUCTION TO SOLID OXIDE ELECTROLYSIS CELLS

High-temperature solid oxide electrolysis cells (SOECs) can simply be considered as the reverse operation of SOFCs. H_2O and CO_2 are introduced into the cathode side of an SOEC (the Ni-based cermet electrode, which is the anode of SOFC) where reduction reactions take place to produce H_2 and CO excessive oxygen in the reactants is pumped to the anode. The whole process is driven by an external DC power supply. Therefore, the term "reversible SOFC" can refer to SOFCs that are capable of being operated as both SOFCs and SOECs, being able to achieve power generation and energy storage in a single device. Under certain circumstances, the combination of SOFCs and SOECs is referred to as SOC technology.

Stable operations of SOEC stacks have been reported by several labs. For example, an SOEC stack was operated at Julich at 800°C with a current density of 0.5 A/cm^2 for over 20,000 h [5], and the degradation rate of the stack during operation was evaluated. In this case, the area specific resistance (ASR) of the SOEC increased at a rate of 3% (1000 h)$^{-1}$, while the voltage applied to the stack increased at a rate of 0.4% (1000 h)$^{-1}$ in order to maintain a constant current. Degradation mechanisms of this SOEC stack are still under investigation. SOEC systems are also commercially available [6]. This commercially available SOEC system uses water steam to generate hydrogen. The rated power input for this SOEC product is 150 kW$_{el}$, generating highly purified H_2 at a rate of 40 Nm^3/h, and reaching an efficiency of 82% based on the LHV of hydrogen.

Operation and fabrication are more difficult for SOECs than for SOFCs. This is because the electrodes and current collectors of SOEC stacks are working under harsher conditions, not to mention the implicit mechanisms of H_2O electrolysis. Apart from H_2O electrolysis using SOECs, the concept and process of co-electrolysis of H_2O and CO_2 using SOECs is another important aspect of EESC systems. By introducing carbon into the fuel feed, production of hydrocarbons and other chemicals is possible. Integration and control strategies of SOEC systems in a distributed power grid have attracted a great deal of attention in recent years.

In regard to SOEC components, novel proton-conducting electrolyte materials have been developed in recent years. In this type of SOEC (also called H-SOEC), protons (H^+) are transported through the electrolyte instead of O^{2-}. During the operation of H-SOECs, H_2O is supplied to the anode of the electrolysis cell (sometimes noted as fuel electrode). Then H^+, produced by water decomposition, migrates through the proton conducting electrolyte and recombines with the reactive species at cathode (noted as oxygen electrode in some literatures) to produce H_2, hydrocarbon, or even NH_3, depending on the type of reactants fed and catalysts applied to the cell.

1.3   MOLTEN CARBONATE FUEL CELLS

Molten carbonate fuel cells (MCFCs) are another type of typical high-temperature fuel cell technology intensively investigated in the past several decades. Operating at around 600°C, MCFCs employ ceramic matrices (β-alumina) soaked with molten carbonates of alkaline metals as the electrolyte. Operating temperatures of MCFCs can vary with the melting points of bicomponent or tricomponent eutectic carbonate mixtures. MCFCs can also utilize carbon-containing fuels. Several natural gas fed MCFC demonstration power plants producing several hundred kilowatts have been built in Europe, Korea, Japan, and the United States.

During MCFC operation, carbonate ions migrate from the cathode side of the electrolyte to the anode side to oxidize fuels there to produce CO_2 and H_2O. As a result, CO_2 must be supplied to the cathode chambers of MCFCs together with air to make up for the carbonate ion concentrations in the molten carbonate electrolyte. MCFCs can be fed with solid carbon fuels and demonstrate impressive performances. This is because the molten carbonate electrolyte can provide a favorable reaction environment for carbon conversion, in which alkaline metal cations are widely accepted catalysts for carbon oxidation.

1.4   INTRODUCTION TO HIGH-TEMPERATURE BATTERIES

Rechargeable high-temperature batteries can be used in relatively large-scale energy storage applications and can possibly be used in situations associated with electricity load leveling, power quality and peak shaving, space power, etc. The main advantages of high-temperature batteries include (1) potential low costs, (2) longer cycling life, and (3) upscaling to megawatt scale. Thermal management and safety are the major concerns however.

There are many types of high-temperature batteries. A typical high-temperature battery is the sodium-beta battery, which is assembled using liquid sodium/sulfur or sodium/nickel chloride [7]. Another type of high-temperature battery is the molten metal battery fabricated with molten metal as the positive/negative electrodes, and molten salt ($MgCl_2$-KCl-NaCl) as the electrolyte [8]. When the molten metal battery is discharging, the Mg metal is oxidized and Mg^{2+} cations as an oxidation product migrate through the molten salt electrolyte and form Sb-Mg alloys at the positive electrode. Open circuit potential (OCP) of this bimetallic system is around 0.5 V at 973 K [9]. This battery demonstrates an energy density of 1000–1500 mAh/cm² at 973 K, depending on current densities ranging from 200–50 mA/cm². A round-trip efficiency of 69% can be achieved when the battery is operating at a constant current density of 50 mA/cm². In general, the energy capacity of a liquid metal battery can be easily scaled up by enlarging the diameter of the cylindrical battery, which enables energy storage on a microgrid scale.

Reversible SOFCs can be combined with metal oxide redox couples to achieve better energy storage performances. Zhao et al. [10] fabricated a solid oxide metal–air redox battery (SOMARB) based on an Fe–Fe_3O_4 redox couple and a reversible SOFC. During the discharge of SOMARB, Fe metals sealed in the anode chamber are oxidized by steam to generate H_2, and the H_2 produced is then used as fuel for the SOFC to generate power. During the charging process, the reversible SOFC acted as an SOEC, and H_2O in the system was decomposed to H_2, which then reduced the metal oxide. Iron beds were integrated into a tubular SOFC to achieve better responses to load variations in the microgrids [11]. The important advantages of SOMARB reaction systems when compared to its conventional counterparts, as the inventors claim [12], are as follows: (1) the solid-to-solid electrochemical reactive interface in SOMARBs is more stable than a solid–liquid one; (2) a SOMARB is capable of operating at high charge and discharge rates as the volume changes induced by the redox processes in the metal do not affect the electrical functionality of the fuel cell; (3) the reactants and products at the oxygen electrode side of the SOMARB are in a dilute phase, preventing possible clogging of the electrode by reaction products in their condensed phases; and (4) the round-trip efficiency of a SOMARB could reach as high as 76.3%.

In summary, a large number of countries and research organizations are leading the research and development of high-temperature EESC technologies, including fuel cells, electrolysis cells, and batteries. Driving forces behind the development of these technologies are substantial, including overcoming the technological challenges in energy security, environmental pollution, and sustainable development. The main obstacles toward the commercialization of these technologies are the cost and technology maturity. Great efforts will be needed to achieve reductions in system costs and promotions of system reliability and stability, as well as identifications of niche markets.

REFERENCES

1. Department of Energy and Environment Mitsubishi Heavy Industries Group, Department of New Energy Business Promotion Mitsubishi Heavy Industries Group. 2013. 4000 h of continuous operation of pressurized SOFC-micro turbine hybrid power system is achived for the first time around the world (in Japanese), September 20, 2013. http://www.mhi.co.jp/news/story/1309205422.html.
2. Kobayashi Y. et al. 2015. Development of next-generation large-scale SOFC toward realization of a hydrogen society. *Mitsubishi Heavy Industries Technical Review* 52: 111–116.
3. Gao Z., Mogni L.V., Miller E.C., Railsback J.G., Barnett S.A. 2016. A perspective on low-temperature solid oxide fuel cells. *Energy & Environmental Science* 9: 1602–1644.
4. Brett D.J.L., Atkinson A., Brandon N.P., Skinner S.J. 2008. Intermediate temperature solid oxide fuel cells. *Chemical Society Reviews* 37: 1568–1578.
5. Fang Q., Blum L., Menzler N.H., Stolten D. 2017. Solid oxide electrolyzer stack with 20,000 h of operation. In *15th International Symposium on Solid Oxide Fuel Cells (SOFC-XV)*. Hollywood, FL.
6. Sunfire. Low-cost hydrogen production. http://www.sunfire.de/en/products-technology/hydrogen-generator, March 8, 2017.
7. Linden D., Reddy T.B. 2001. *Handbook of Batteries*, 3rd edn. McGraw-Hill, Inc, NY.
8. Bradwell D.J., Kim H., Sirk A.H.C., Sadoway D.R. 2012. Magnesium–antimony liquid metal battery for stationary energy storage. *Journal of the American Chemical Society* 134: 1895–1897.
9. Bradwell D.J. 2011. *Liquid Metal Batteries: Ambipolar Electrolysis and Alkaline Earth Electroalloying Cells*. Massachusetts Institute of Technology, Boston, MA.

10. Zhao X., Gong Y., Li X., Xu N., Huang K. 2013. Performance of solid oxide iron-air battery operated at 550°C. *Journal of the Electrochemical Society* 160: A1241–A1247.
11. Zhang C. et al. 2016. A dynamic solid oxide fuel cell empowered by the built-in iron-bed solid fuel. *Energy & Environmental Science* 9: 3746–3753.
12. Zhao X., Li X., Gong Y., Huang K. 2014. Enhanced reversibility and durability of a solid oxide Fe–air redox battery by carbothermic reaction derived energy storage materials. *Chemical Communications* 50: 623–625.

2 Solid Oxide Fuel Cells

2.1 INTRODUCTION

High-temperature solid oxide fuel cells (SOFCs) facilitate more efficient and lower/zero pollution transformation of fuels into electricity as compared to traditional power generation technologies. With increasing fuel prices and tighter emission regulations, SOFCs are becoming increasingly more attractive due to their high fuel flexibility for hydrocarbons, reforming gas, biogas, and coal syngas. This is a critical advantage over other low-temperature fuel cells that require hydrogen as a fuel. The fuels used in SOFCs are relatively cheap, safe, and more readily available than hydrogen, allowing for lower operational costs and higher commercialization potential. In addition, since fuels in SOFCs do not contact oxidants directly as in traditional combustion processes, high CO_2 concentration streams can be obtained at the anode side, which is beneficial for CO_2 sequestration [1–4].

SOFC stacks can adopt different geometries. For example, planar SOFCs are widely employed due to their relatively high power densities, while tubular SOFCs are developed for their high sealing feasibilities. Planar tube SOFCs are fabricated to combine both advantages. There are other fuel cell designs based on the shape of current collectors, and these will be discussed in the latter part of this chapter.

Despite the many SOFC products coming onto the market, the fuels that are being used in these systems are restricted to small molecules such as H_2, reforming gas, coal gas, and nature gas. Promising anode catalysts and anode designs for heavier hydrocarbon fuels have yet to be fully developed or applied. Thus, the long-term operation of SOFCs using larger carbonaceous fuels is still impeded by carbon deposition (residue of hydrocarbon conversion) and anode degradation. Sealing and stack fabrication technology are also two factors hindering the scaling up of SOFC systems.

Apart from fuel cell stack related issues, focus on technical challenges involving balance of plant (BoP) equipment is also needed. Desulfurization of city gas, water treatment for steam–methane reforming, design of flue gas burner, and compact heat exchanger for heat recovery are key BoP equipments in SOFC-based power systems. As identified, poor design of BoP equipment or subsystems is the main reason for SOFC system failure.

Currently, SOFCs are still at the early stages of commercialization. However, several companies in the United States, Australia, Japan, and Europe have constructed well-designed SOFC systems that have been well received by customers.

2.2 ELECTROCHEMISTRY OF SOFCs

The positive electrode–electrolyte–negative electrode assembly (PEN) in an SOFC is the main location for electrochemical reactions. A geometrical sketch of a PEN is shown in Figure 2.1. During SOFC operations, gaseous oxygen is reduced to oxygen ions (O^{2-}) in the cathode region of the PEN. Then the generated O^{2-} migrates through the ionic conducting electrolyte and reaches the anode side of the electrolyte. The fuel, or other reducing species at the anode region, is then oxidized by O^{2-} and releases electrons to the external circuit. The electrochemical reaction at the cathode can be expressed as Equation 2.1:

$$O_2 + 4e^- \rightarrow 2O^{2-} \tag{2.1}$$

The reversible cell voltage, more commonly referred to as the open circuit voltage (OCV), is determined by the chemical potential difference between the anode and cathode regions. OCV values of a given SOFC reaction system at a steady state can be evaluated using the Nernst equation, shown as Equation 2.2 and Equation 2.3:

$$E_{rev} = \frac{RT}{4F} \ln\left(\frac{p_{O_2,\text{cathode}}}{p_{O_2,\text{anode}}} \right) \tag{2.2}$$

$$E_{rev} = -\frac{\Delta G^0}{n_e F} - \frac{RT}{n_e F} \ln p_i^{v_i} \tag{2.3}$$

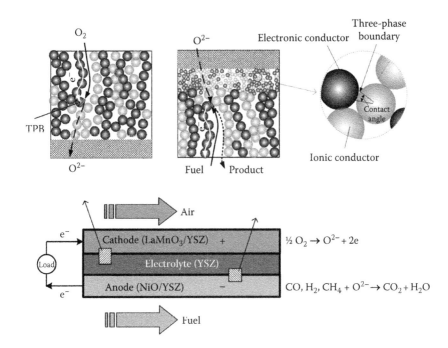

FIGURE 2.1　Basic operating principles of SOFCs.

where

$p_{O_2,cathode}$ and $p_{O_2,anode}$ are the oxygen partial pressures at the cathode and anode
F is Faraday's constant
ΔG^0 is Gibbs free energy change
n_e is the number of electrons transferred during the electrochemical reaction
T is the cell operating temperature
p_i is the partial pressure of the different species

The partial pressure of oxygen at the anode is determined by the thermodynamic equilibrium of the oxidation reaction. The SOFC anode reactions of several typical fuels (H_2, CO, CH_4, and C) are expressed in Equations 2.4 through 2.7:

$$H_2 + O^{2-} \rightarrow H_2O + 2e^- \tag{2.4}$$

$$CO + O^{2-} \rightarrow CO_2 + 2e^- \tag{2.5}$$

$$CH_4 + 4O^{2-} \rightarrow CO_2 + 2H_2O + 8e^- \tag{2.6}$$

$$C + 2O^{2-} \rightarrow CO_2 + 4e^- \tag{2.7}$$

Figure 2.2 shows the calculated thermodynamic OCV values of different fuels as a function of operation temperatures. The results are calculated based on fuel–air chemistry.

For a given working condition, if a certain current is produced by the cell, the cell potential of the SOFC will be lower than the OCV value. This decrease in voltage is

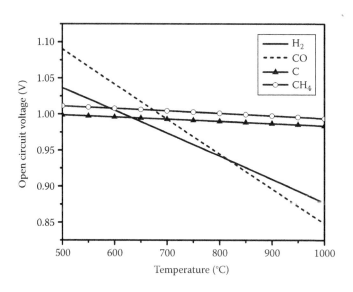

FIGURE 2.2 Open circuit voltages for four typical fuels at different temperatures.

mainly attributed to several types of voltage loss, including (1) ohmic polarization loss due to finite electronic and ionic conductivity, (2) activation polarization loss due to reaction barriers, and (3) concentration polarization loss due to the consumption of reactants and finite mass transport rates.

SOFC design and operation should be able to minimize the three types of polarization losses mentioned earlier. The main functions of the electrolyte are ionic conduction and electron transport blocking at SOFC operating temperatures. The anode and cathode should be able to facilitate both ionic and electronic transport and should have high activities for fuel oxidation and oxygen reduction. The microstructures of the electrodes need to be designed properly and fabricated to meet the requirements of mass transport and charge transfer, while at the same time achieve a higher density of electrochemical reactive sites [5,6]. Apart from physical or chemical properties of single materials, mechanical and chemical compatibility of SOFC components also needs to be considered. From a mechanical point of view, coefficients of thermal expansion (CTE) in fuel cell components should be similar to each other in order to avoid CTE mismatches between adjacent materials, which can cause PEN structures to collapse during SOFC operation. Some perovskite materials can react with zirconia-based electrolyte materials through solid-phase reactions during the sintering process at elevated temperatures, forming an isolating layer between the electrolyte and electrode, thus hindering electrochemical reactions at the electrode.

Yttria-stabilized zirconia (YSZ) is a common electrolyte material that exhibits acceptable ionic conductivity at 650°C–1000°C [7]. In addition to zirconia- or ceria (gadolinium-doped ceria, GDC)-based materials, many types of ionic conductors have been studied, e.g., $LaGaO_3$-based perovskites [8], pyrochlores $(Gd,Ca)_2Ti_2O_{7-\delta}$, and $Ln_{10-x}Si_6O_{26\pm\delta}$ derived apatite materials [9]. Additionally, proton-conducting materials based on ceria and barium oxides have also been developed, such as H-SOFC [10].

Cathodes of SOFCs are usually made of mixed conducting perovskite-type ceramic materials. Lanthanum strontium manganite (LSM), lanthanum strontium cobalt ferrite (LSCF), or lanthanum strontium ferrite (LSF) [11] are commonly used.

Anodes of SOFCs are usually made with nickel YSZ. Nickel is highly active toward the catalytical cracking of hydrocarbon fuels. Nickel is also electronically conductive and stable under reducing atmospheres. YSZ in the anode cermet serves as an ionic conductor that can effectively extend the triple-phase boundary (TPB), as well as serve as a sintering inhibitor of nickel catalysts [12,13]. Cu-based, Co-based, and ceria (CeO_2)-based anodes have also been developed and reported as electrodes for hydrocarbon conversions [14]. Despite other advanced anode materials intensively reported in literature, Ni-YSZ anodes are commonly used as anode materials.

2.3   SOLID OXIDE FUEL CELLS FUELING WITH SYNGAS AND HYDROCARBONS

Reaction mechanisms in SOFC anodes can be extremely complicated. Elementary reaction steps include adsorption of reactive species, dissociation of reactants, and mass transport and charge transfer reactions. Therefore, a deep understanding of reaction mechanisms and identifying rate-determining steps of the fuel conversion process are of great importance. A detailed fuel conversion mechanism will provide guidance for the design and optimization of SOFCs.

2.3.1 Hydrogen Electrochemical Oxidation

As discussed earlier, Ni-YSZ cermet anodes have been intensively investigated in SOFCs. In general, the porous structures of these anodes allow for much larger densities of electrochemically reactive interfaces and sufficient passways for mass transport. However, due to the complexity of the anode cermet microstructures, it is hard to establish a quantitative description of the electrode microstructures (the area of electrochemically reactive interface cannot be easily calculated). Therefore, to simplify the microstructures of anodes and the mass transport process, studies focusing on anode reaction mechanisms often adopt pointed or patterned Ni anodes [15,16]. Adsorption and reactive area lengths of patterned electrodes can be easily determined because of their regulated geometry. Three reaction mechanisms of H_2 oxidation are summarized here:

1. *Hydrogen spillover.* Hydrogen spillover starts from the adsorption and dissociation of hydrogen. Mass transport and charge transfer at the reactive area and in the bulk of YSZ are also considered [17]. The main reaction steps include (1) H_2 adsorption on the Ni surface of the anode; (2) ionization of the absorbed species to form proton; (3) the proton then migrates from the adsorption site to the electrochemically reactive interface, and reacts with the oxygen ion (O^{2-}) transported through the electrolyte to form the hydroxyl ion and other important reactive species. In addition, the protons formed may migrate into the bulk of the YSZ and then react with the lattice oxygen ions; and (4) the produced H_2O then escapes from the system. This mechanism was modified based on the assumption that the hydroxyl ions formed at the YSZ surface can migrate to the reactive surface [18]. This modified mechanism has been widely accepted. Goodwin et al. [19] and Zhu et al. [20] have proposed a thermodynamic and kinetic database for H_2 electrochemical oxidation on Ni surfaces.
2. *Oxygen spillover.* The oxygen spillover mechanism was proposed by Mizusaki et al. [21], oxygen ions are assumed to be transported via spillover process from the electrolyte to the reactive area. The adsorbed species on the Ni surface are assumed to include O(Ni), OH(Ni), H(Ni), and H_2O(Ni). In this mechanism, three possible rate-determining steps are proposed: (1) adsorption and desorption of H_2 and H_2O, (2) surface diffusion of adsorbed species, and (3) exchange of adsorbed species across the reactive area.
3. *Interstitial hydrogen.* In the interstitial hydrogen mechanism, the elementary reaction mechanisms include the following steps: (1) reactive species diffusion, (2) dissociative adsorption on Ni and YSZ surfaces, (3) surface diffusion process of surface elements, and (4) species charge transfer and H_2O formation.

It is widely accepted that surface diffusion plays a key role in affecting the electrochemical reaction kinetics at the reactive zone. Due to the limited rate of surface diffusion, effective electrochemically active area are confined to narrow strips (with widths of 100 nm) along the reactive zone. The possible reaction mechanisms of the competitive adsorption and the following surface diffusion processes are regarded as possible limiting steps.

2.3.2 CARBON MONOXIDE ELECTROCHEMICAL OXIDATION

Carbon monoxide, found in city gas, syngas, and reformed gas, is a widely used fuel for SOFCs. It is also an important intermediate gas content for the direct conversion of carbonaceous fuels in SOFCs. A wide variety of experiments under different operation conditions using different material systems have been conducted to investigate the electrochemical oxidation of CO in SOFCs.

Tests of CO electrochemical oxidation have been carried out on different electrode systems. Ni patterned anodes, Ni point anodes, and Ni/YSZ porous anodes are commonly investigated electrochemical systems for CO digestion. The electrochemical reaction rate of CO is commonly considered to be lower than that of H_2 [22]. According to tests on a patterned anode by Sukeshini et al. [23], the peak power density using CO as a fuel is about 50% of that when using H_2 under the same temperature, as shown in Figure 2.3b.

Matsuzaki et al. [24] tested the reaction rates of H_2 and CO, and found that the electrochemical oxidation rate of H_2 was 1.9–2.3 times and 2.3–3.1 times higher than that of CO at 1023 and 1273 K, respectively [25,26].

2.3.3 SOFC PERFORMANCE AND MECHANISMS WITH H_2/CO MIXTURE FUEL

When CO and H_2 coexist, the situation becomes more complex due to the interactions between CO, H_2, and their products. Jiang et al. [27] reported that when fueled with H_2 and CO, cell performances were close to that of pure H_2 when the CO molar fraction was at 55 mol%. As long as the H_2 content in the fuel stream was higher than 50 mol%, high cell performances close to pure H_2 operation can be achieved because most of the CO is converted to H_2 before being oxidized at the anode. Costa-Nunes et al. [28] compared cell performances when fueled with H_2, CO, and syngas and predicted that the higher H_2 activity is attributed to hydrogen spillover from the Ni to the YSZ surface. In contrast, CO adsorbed on the metal can hardly spill over towards the oxide support. Habibzade [29] tested CO and H_2-fueled cell performances on patterned Ni anodes and developed a detailed electrochemical model using a state-space modeling approach based on the MATLAB platform and linked it to CANTERA to conduct electrochemical calculations. The elementary reactions used in this study are shown in Figure 2.4.

2.3.4 HYDROCARBON ELECTROCHEMICAL OXIDATION

Hydrocarbons such as CH_4, C_3H_8, and C_4H_{10} are another type of fuel commonly used. Propane (C_3H_8) and butane (C_4H_{10}) are ideal energy storage media for potable energy supplies. Apart from using hydrocarbons in SOFCs with external or internal

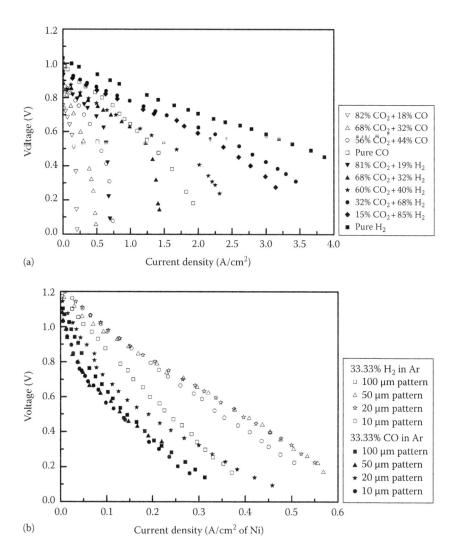

FIGURE 2.3   SOFC performance with CO fuel on different electrode systems (a) porous anode (b) patterned anode. (From Sukeshini, A.M. et al., *J. Electrochem. Soc.*, 153, A705, 2006.)

reforming processes (steam or CO_2 is required to turn hydrocarbon into CO and H_2), researchers have developed dry-hydrocarbon-fueled SOFC anodes to implement hydrocarbon direct conversion, which refers to the direct electrochemical oxidation of hydrocarbons or the cracking of species.

Murray et al. [30] studied CH_4 electrochemical oxidation in SOFCs. The power density in their study reached 0.37 W/cm^2 at $650°C$. This is close to the cell performances when fueled with hydrogen. They suggested that direct methane oxidation is the main reaction. Although the H_2O and CO_2 produced might be beneficial to CH_4 reforming, the rates of methane reforming were slow compared to CH_4 direct

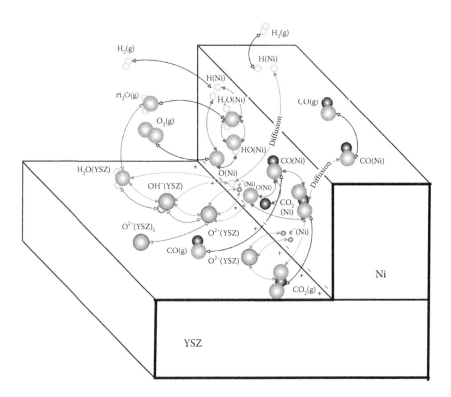

FIGURE 2.4 Schematic of the electrochemical oxidation processes of CO and H_2 in patterned Ni/YSZ anode.

electrochemical oxidation. The different shapes of the electrochemical impedance spectra (EIS) when fueled with dry/humidified methane are more evident, suggesting the direct digest of CH_4. The differences in EIS results suggest that performance of dry-CH_4-fueled fuel cells is lower than those fueled by H_2 as a consequence of higher electrode impedance for methane oxidation

Putna et al. [31] and Park et al. [32] suggested that the direct oxidation of CH_4 and C_4H_{10} is performed through the following reactions:

$$CH_4 + 4O^{2-} \rightarrow CO_2 + 2H_2O + 8e^-$$ (2.8)

$$C_4H_{10} + 13O^{2-} \rightarrow 4CO_2 + 5H_2O + 26e^-$$ (2.9)

Mogenson et al. [17] comprehensively reviewed hydrocarbon conversions in SOFCs and suggested that the direct electrochemical oxidation reactions of hydrocarbon fuels are highly unlikely to occur in one step and may contain the following elementary reaction steps:

$$CH_4 + O^{2-} \rightarrow CH_3OH + 2e^-$$ (2.10)

$$CH_3OH + 2O^{2-} \rightarrow HCOOH + H_2O + 4e^- \qquad (2.11)$$

$$HCOOH + O^{2-} \rightarrow CO_2 + H_2O + 2e^- \qquad (2.12)$$

Meanwhile, the cracking of hydrocarbons and electrochemical reactions of cracking products can also be possible:

$$C_xH_y \rightarrow xC + y/2H_2 \qquad (2.13)$$

$$C + 2O^{2-} \rightarrow CO_2 + 4e^- \qquad (2.14)$$

It's commonly accepted that, different anode materials lead to different reaction mechanisms and show different catalytic activities, which will significantly affect the anode designs for targeted reactants and products. In order to improve fuel cell performance and stability, it is important to modify anode materials based on the demands of the fuel conversion.

2.3.5 CARBON DEPOSITION

Carbon decomposition is a common phenomenon observed by researchers at normal SOFC operating temperature ranges (600°C–900°C). Although anode conductivity can be increased to some extent from the formation of conductive graphite through carbon deposition, this layer of carbon may cause the collapse of Ni–YSZ anode structures, resulting in the loss of the catalyst. Therefore, carbon deposition in SOFC anodes needs to be suppressed. To reduce carbon deposition, advanced anode materials, novel operation strategies, and new fuel cell designs have been proposed in the last several decades. The amount of carbon deposited at the anode is affected by steam/carbon ratios, temperatures, types of anode materials, and operating current densities.

Commonly, there are three types of observed deposited carbon according to researchers and are classified by shape: (1) fibrous carbon, (2) encapsulating carbon, and (3) pyrolytic carbon [33]. At high operating temperatures, fibrous carbon and pyrolytic carbon are the predominant types of carbon deposition. The appearance of fibrous carbon and pyrolytic carbon was shown by Anderson et al. [33]. Mechanisms of the formation of these two types of carbon have been studied, where the pyrolytic carbon results from gas-phase deposition reactions [34–37].

Besenbacher et al. [38] suggested that adding a trace amount of gold in nickel catalysts can release carbon depositions significantly. Rostrup-Nielsen et al. [39] suggested that Au can inhibit the dissociation of methane, based on the results of their density functional theory modeling. Calculation results based on the DFT model were confirmed through experimental observations conducted by Triantafyllopoulos et al. [40]. Carbon deposition process on Ni (1 at% Au added)/YSZ catalysts was found largely suppressed under methane atmosphere.

2.4 MODELING AND SIMULATION OF SOLID OXIDE FUEL CELLS

Because experimental studies on SOFCs are normally expensive, time-consuming, and labor-intensive, it is important to develop theoretically model and predict the detailed processes of the reactions and transports, so as to provide useful information

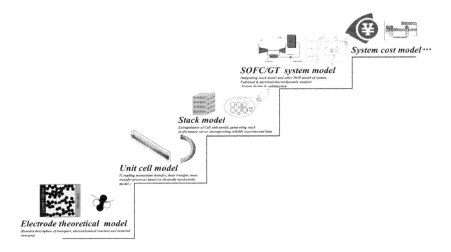

FIGURE 2.5 SOFC modeling: from electrode reaction mechanism to unit, stack, and systems.

to guide experiments. Combination of experiment and modeling saves both time and cost. When hydrocarbon or syngas is fed as fuel, the reaction and transport processes become extremely complex. The processes may include gas phase transport, porous media molecular diffusion and Knudsen diffusion, adsorption/desorption, surface dissociation, surface diffusion, surface reaction, as well as charge transfer reactions. Detailed models that take elementary homogeneous/heterogeneous thermal chemistry, electrochemistry, and coupled transport processes into consideration are still lacking and require further development.

Numerous mathematical models of different scales have been developed, starting from the molecular level, electrode level, single-cell level, and up to stack and system levels, as shown in Figure 2.5. Many researchers have developed micromodels to describe the charge transfer processes near the three-phase boundary [11,41–43]. These models usually consider the electrode layer as a discrete volume occupied by randomly packed spheres of ionic and electronic conductive materials. Other detailed models deal with surface reactions and the diffusion of adsorbed species, and charge transfer reactions. In generally, researchers have developed porous composite electrodes while at the same time considering gaseous species transport through the porous phase and Ohm's law was adopted for both voltage and current calculation [44,45]. These models are formulated based on a continuum differential-equation setting and usually in one spatial dimension through the electrode thickness. Stack-level models are usually based on transport of momentum (fluid dynamics), transport of energy (conductive, convective and radiative heat transfer), and spices transport. Electrochemical and chemical reactions are also coupled to the fluid field. These models are useful in stack design and optimization [46,47]. Such models accommodate complex flow configurations and cell layouts to predict electrochemical performances, temperature distributions, pressure drops, and stress distributions.

At the beginning of simulations, button SOFCs are normally used to validate electrochemical models and model parameters. The determined parameters are then extended to unit cell models and stack models by adding overall mass balance, charge balance, energy balance, as well as momentum balance into the models. During model calculation, three commonflow field configurations (co-flow, counterflow, and cross-flow) are simulated and compared. Due to the computing complexity of SOFC stacks, unit cell models can be considered as the foundation of stack model debugging and can also be used to check simulation rationality.

2.4.1 PEN Modeling of Solid Oxide Fuel Cells

An accurate SOFC PEN submodel is important for understanding the reaction/transport processes in a real unit cell/stack. Thus, during SOFC PEN submodel development, researcher have to meet two criteria: (1) the PEN submodel has to predict the cell performance correctly and should became a stand-alone model when proper boundary conditions are applied; (2) the PEN submodel has to be feasible enough to be integrated into multidimensional CFD codes for SOFC stack simulations.

Apart from chemical and electrochemical kinetic parameters, parameters related to the microstructure of different cell layers are crucial for model accuracy. These parameters are electrochemical reactive area, cell layer thickness, ionic and electronic conductivity, pore size, porosity and tortuosity of electrodes. Density of electrochemical reactive area in the electrode is evaluated based on the percolation theory. The electrode consisting both ionic conducting and electronic conducting phases is simplified as random packing of binary spheres. The simplification makes it easier to determine the coordinate number for each sphere, which is required during the calculation of percolation theory. The pore size and porosity of electrodes can be characterized based on quantitative stereology using mercury porosimetry as well as image processing based on SEM images. Meanwhile, the surface diffusion process can be added into the model by extending transport terms in elementary surface reaction kinetic equations or by modifying the proposed diffusion circuit analog.

As mentioned earlier, SOFC button cells are widely used to study the anodic and cathodic processes under various operating conditions. Since button cells are usually small (e.g., 26 mm in diameter), the effects of nonuniformity in temperature and velocity fields on cell performances can be neglected. Thus, button cell, experiments are valuable for PEN submodel calibrations and validation to ensure the feasibility of SOFC PEN submodels.

2.4.2 Cell Unit Modeling of Solid Oxide Fuel Cells

Cell unit modeling of SOFCs is a tool for the optimization of cell geometry. For example, Figure 2.6 shows the cell unit modeling of a planar SOFC unit. Planar SOFC unit models can be developed by utilizing continuum electrode models to describe electrode statistical properties and can be further coupled with transport governing equations for charge/mass/momentum/energy. The calculation domain is chosen based on a type of planar SOFC design with a flow channel length of 100 mm, a channel height of 1 mm, and an interconnect height of 2.5 mm (1.5 mm without rib).

In this unit cell model, cross sectional distribution of oxygen concentration inside the cathode can be calculated and visualized. It is obvious in Figure 2.7a that

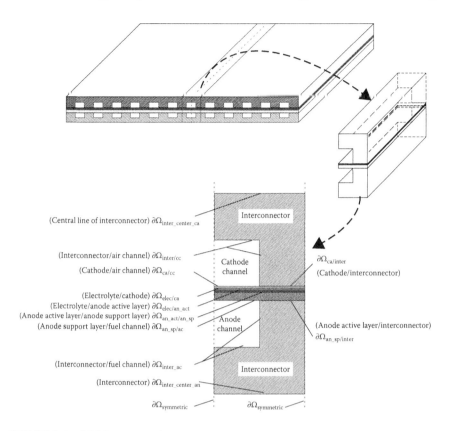

FIGURE 2.6   SOFC unit modeling in rib cross section.

oxygen concentration under the rib are much lower than that in the channel under both operation voltages (0.8 V and 0.4 V). In this situation, oxygen diffusion is the limiting step of the cell operation. Figure 2.7b shows cross sectional hydrogen concentration distribution inside anode of the SOFC unit, both results calculated under 0.8 V and 0.4 V show a nonuniform distribution of hydrogen content. Figure 2.7c shows current density distributions under the rib and under the channel at 0.8 and 0.4 V, respectively, with different rib widths. The results suggest that reducing the rib width can effectively promote oxygen transport and increase current density. However, smaller ribs will also lead to larger ohmic resistances. The optimization of the rib and channel widths requires comprehensive consideration of the effects of electronic/ionic conduction, reaction, and transport processes.

Another example is a three-dimensional comprehensive model of an anode-supported SOFC unit [48]. This model takes of mass/heat/momentum transport, chemical/electrochemical reactions, and charge transport into consideration to simulate the nonuniform distribution of fluid velocity in an SOFC unit. In this model, the effects of fuel velocity nonuniformity, working voltage, fuel flow rate, flow pattern, and fuel type on the fuel concentration differences among the flow channels are investigated. Figure 2.8 shows the calculated results of fuel flow non-uniformity.

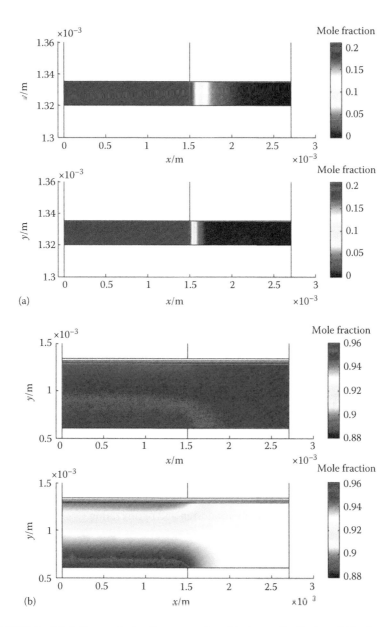

FIGURE 2.7 Modeling results in rib cross section: (a) air gas distribution, (b) hydrogen gas distribution. (*Continued*)

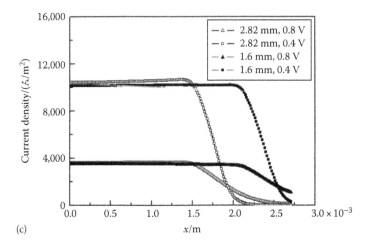

(c)

FIGURE 2.7 (Continued)   Modeling results in rib cross section: (c) effects of rib width and operating voltage.

2.4.3 Solid Oxide Fuel Cell Stack Modeling

To obtain the detailed distribution of current on each cell within a planar SOFC stack, the effects of reactant flow and temperature distributions inside to stack have to be comprehensively considered. The model includes all the effects of heat, mass transport, and manifold geometry.

Figure 2.9 demonstrates Z-type gas supply in an SOFC stack. Figure 2.10a provides the distribution of fuel in different cells is almost uniform with increasing velocity from top to bottom. This implies that cells in the top of the stack are supplied with less fuel. Fuel acceleration along the channels in each cell, mainly due to heat effects, can also be observed. As shown in Figure 2.10b, cell temperatures in the middle of stacks are higher than those in the top or bottom sections. This is a consequence of the isothermal thermal boundary conditions. Figure 2.10c further proves that hydrogen mole fraction distributions among cells are nonuniform, which implies that fuel utilization among cells is different. These phenomena demonstrate the importance of stack-level geometry design and optimization.

2.5 SOLID OXIDE FUEL CELL SYSTEM

2.5.1 Typical System Configurations

Design of SOFC power generation systems involves the integration of different power-generating devices and the adjustment of system parameters based on the demands of different applications. Design of an SOFC-involved power system should be based on careful considerations and detailed calculations. Necessary considerations during system design and integration include fuel acquisition and processing, SOFC stack design, heat recovery, and other BoP components.

SOFC/gas turbine (GT) hybrid systems are one of the most promising technologies for SOFC-based power systems. In SOFC/GT systems, an SOFC unit is used as a top cycle so that the latent heat in its high-temperature exhaust gases can be further utilized by a bottom cycle (gas turbine). SOFC/GT hybrid systems can be roughly classified into two types as shown in Figure 2.11. The system configuration in Figure 2.11a is defined as a direct hybrid cycle, in which flue gas from an SOFC stack goes into the combustion chamber of a gas turbine along with a certain amount

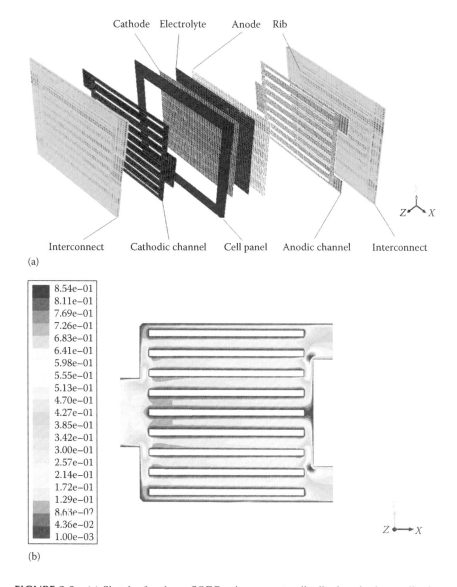

FIGURE 2.8   (a) Sketch of a planar SOFC unit: parameter distributions in the anodic channel; (b) velocity field of hydrogen.   (*Continued*)

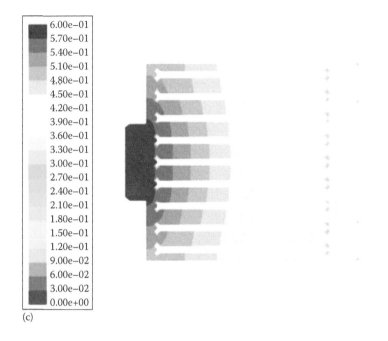

(c)

FIGURE 2.8 (*Continued*)   (c) Mole fraction of hydrogen. (Reprinted from *Int. J. Hydrogen Energy*, 40, Lin, B., Shi, Y., Ni, M., and Cai, N., Numerical investigation on impacts on fuel velocity distribution non-uniformity among solid oxide fuel cell unit channels, 3035–3047, Copyright 2015, with permission from Elsevier.)

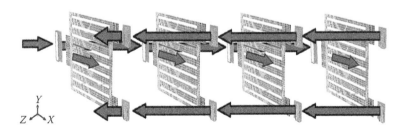

FIGURE 2.9   Sketch of the SOFC stack gas supply scheme. (Reprinted from *Appl. Thermal Eng.*, 114, Lin, B., Shi, Y., and Cai, N., Numerical simulation of cell-to-cell performance variations within a syngas-fueled planar solid oxide fuel cell stack, 653–662, Copyright 2017, with permission from Elsevier.)

of additional fuel to fully extract the chemical energy in the flue gas. High system efficiency can be achieved using this direct integration of SOFC and gas turbine. The system configuration in Figure 2.11b is defined as an indirect hybrid cycle. In this design, flue gas from an SOFC stack is led to a recuperator to heat air from a compressor; the heated air then expands in a gas turbine to generate power. In this configuration, the flue gas of the SOFC is not included as the working fluid of the gas turbine. This configuration allows the SOFC and gas turbine to be operated at different pressures, which would effectively reduce the complexity of the whole system, save the cost of SOFC materials, and lower the sealing requirements.

As a typical example of direct SOFC/GT hybrid systems, Siemens–Westinghouse developed the world's first 220 kW SOFC/GT hybrid system that comprises a 180 kW SOFC stack and a 40 kW micro gas turbine as shown in Figure 2.12. Here,

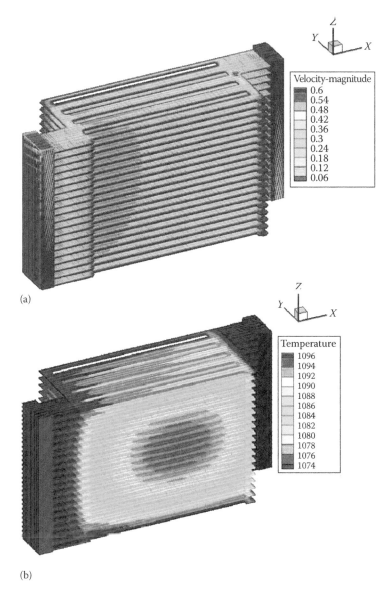

(a)

(b)

FIGURE 2.10 Parameter distributions within a 20-cell SOFC stack: (a) fuel velocity, (b) temperature.   (*Continued*)

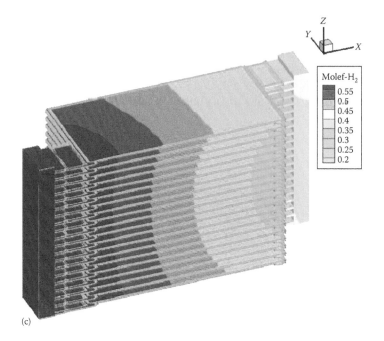

(c)

FIGURE 2.10 (*Continued*)   Parameter distributions within a 20-cell SOFC stack: (c) hydro-gen mole fraction. (Reprinted from *Appl. Thermal Eng.*, 114, Lin, B., Shi, Y., and Cai, N., Numerical simulation of cell-to-cell performance variations within a syngas-fueled planar solid oxide fuel cell stack, 653–662, Copyright 2017, with permission from Elsevier.)

high-temperature exhaust gases from the SOFC stack are introduced into the micro gas turbine to generate electricity [50].

Operation temperatures of the SOFC stack can be as high as 1000°C. Desulfurized natural gas is supplied to the SOFC stack using a fuel compressor. Part of the depleted anode gases is then recirculated to a reformer to supply heat and steam required by the CH_4 reforming process, while the rest of depleted gases flow into the combustion chamber of the gas turbine. In the combustion chamber, the depleted gases from the anode are mixed with residue air from the cathode to produce heat for gas turbine power generation. The system proof-of-concept was tested at the National Fuel Cell Research Center, University of California, Irvine for about 3000 h, and the achieved a unit electrical efficiency of 52% (LHV) [51].

As an example of indirect SOFC/GT systems, a conceptual design of an SOFC/GT system consisting of a microturbine and a nonpressurized fuel cell is shown in Figure 2.13 [52]. This design is a combination of a fuel cell, a gas turbine, and a recuperator. Flow and heat requirements of the microturbine and SOFC have been well matched, leading to a highly integrated package. Compressed air is heated in a high-temperature recuperator using exhaust gas from the SOFC module. The hot compressed air then expands through the turbine and generates power, and drives the compressor.

SOFC hybrid systems can also be fueled using coal-derived syngases. As shown in Figure 2.14, the DOE (United States) Federal Energy Technology Center planned

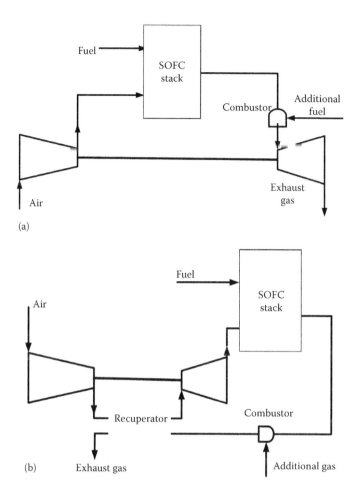

FIGURE 2.11   Direct and indirect integration of SOFC and gas turbine: (a) direct hybrid cycle, (b) indirect hybrid cycle.

a coal-based SOFC power system [3]. This system was designed based on Siemens–Westinghouse tubular SOFC stacks. The system also consists of a Destec coal gasifier, a reheat gas turbine, a heat recover turbine bottom cycle, a catalytic oxidizer, and a membrane-based H_2 and O_2 separator. The designed electrical efficiency is around 60% (LHV). Earlier studies on coal-fueled SOFC/GT power plants were conducted by Lobachyov et al. [53]. This theoretical study linked the Conoco coal gasification process with the SOFC and gas turbine. They reported that the overall electrical efficiency can reach around 60%.

2.5.2   Fuel Processing for SOFC Systems

In regard to the fuels for SOFC systems, the primary fuel composition of coal-derived syngas is similar to that of reformed gas from natural gas. In fact, the hydrogen

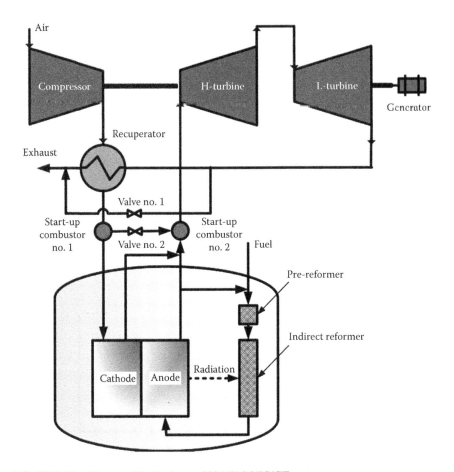

FIGURE 2.12   Siemens–Westinghouse 220 kW SOFC/GT system.

content of reformed gas made from natural gas is higher than that of coal-derived gas, while syngas produced from coal possesses a higher carbon monoxide content. An issue related to coal-based gaseous fuels is mineral contamination, which requires the removal of these contaminants before they enter the fuel cell. The contaminants typically include H_2S, NH_3, HCl, particulates, and tars. Therefore, the cleaning of raw gases requires a heat exchanger, a particle removal unit, an ammonia scrubber, an acid gas scrubber, a sulfur recovery, and polishers.

As mentioned earlier, in an SOFC-based power generation system, fuel-processing subsystems are necessary to convert commercially available gas, liquid, and solid fuels into feedstock for SOFCs that are suitable for fuel cell reactions. The choice of internal/external reforming of fuels, the type of raw fuels, the preheater design, and the recycling of anode gases are key factors that affect fuel cell stack designs. Typically, there are two types of fuel-processing technologies for hydrocarbon-based fuels that are often considered: (1) external reforming and (2) internal reforming. Both technologies will be discussed in the following sections.

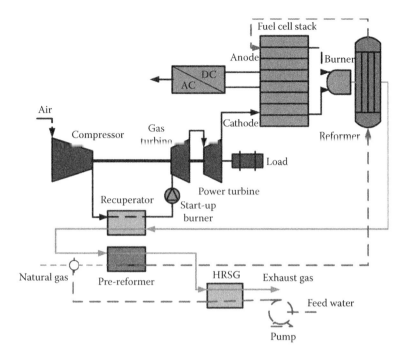

FIGURE 2.13 Process schematic for indirect fuel cell–microturbine hybrid system.

2.5.2.1 Internal Reforming

The internal reforming process achieves reformates in a unit integrated to an SOFC stack rather than in a separate reactor. Thus, internal reforming can effectively reduce system complexity. Hydrocarbon reforming can use heat from the electrochemical reactions. Apart from fuel processing, reforming units can also be regarded as a cooling procedure for SOFC stacks. Moreover, internal reforming is also attractive for faster loading responses [54].

Internal reforming can usually be divided into two types: indirect internal reforming and direct internal reforming. Figure 2.15a shows the schematic of a tubular SOFC stack with an indirect internal reformer and pre-reformer [55]. Here, raw fuel is firstly introduced into the pre-reformer and then goes into the internal reformer. Recycled flue gas from the anode is then mixed with the inlet gas using an ejector to provide steam for the steam–methane reforming process. At the same time, the SOFC stack is cooled by this reforming reaction while the heat generated from the electrochemical reactions in the SOFC stack can serve as a heat source of the steam reforming reaction. Finally, residual fuel gas from the anode is burned in a post combustor to generate high-temperature exhaust gases, which can be further recovered by a gas turbine or a heat recuperator. Results from Brus et al. [56] suggest that radiative heat transfer is important for indirect internal reforming. Figure 2.15b shows a typical indirect internal reformer of a planar SOFC.

For the direct internal reforming process, steam used by the hydrocarbon reforming reaction can be supplied partially through the electrochemical oxidation of hydrogen.

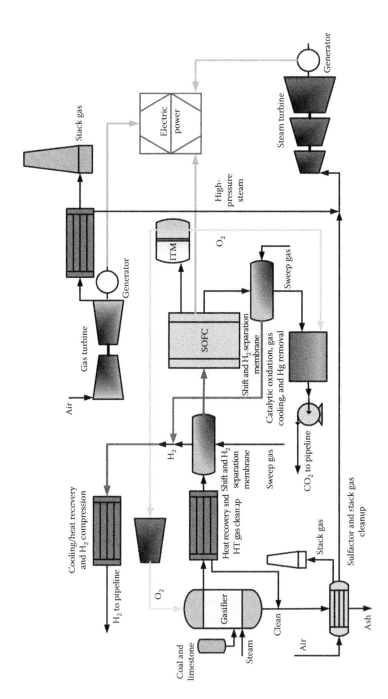

FIGURE 2.14   Demonstration of FutureGen simplified cycle. (With kind permission from Springer Science+Business Media: *Solid State Ionics*, Solid oxide fuel cell technology development in the U.S., 177, 2006, 2039–2044, Williams, M.C., Strakey, J.P., Surdoval, W.A., and Wilson, L.C.)

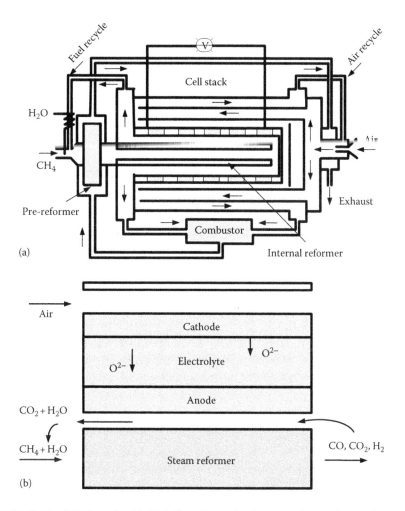

FIGURE 2.15 SOFC stack with (a) indirect internal reformer and pre-reformer for tubular SOFC stack and (b) indirect internal reformer for planar SOFC stack.

Figure 2.16 showed a conceptual design of direct internal reforming for SOFCs by adding a catalytically active layer [57,58]. The added catalyst layer ensures that catalytic reactions and electrochemical reactions are separated on different reactive sites even though both take place inside the anode. This will allow for the individual application of optimized materials for each functional layer. The reforming process is a highly endothermic process however and can lead to a large temperature gradient as well as to large thermal stresses over the anode. In addition, steam reforming requires relatively large amounts of water, which will result in decrease in SOFC open circuit potentials and electrochemical reaction rates.

As mentioned in previous sections, direct electrochemical oxidation of hydrocarbon fuels in SOFCs is reported to be possible and direct power generation based on hydrocarbons in SOFCs will allow for the elimination of fuel reformers, largely

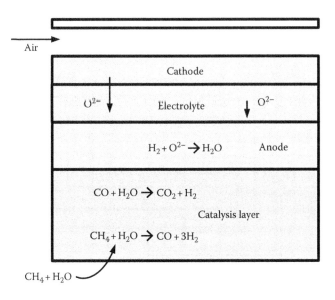

FIGURE 2.16 A conceptual design of direct internal reforming process associated with the anodic catalytic layer. (Reprinted from *J. Power Sources*, 193, Klein, J., Hénault, M., Roux, C., Bultel, Y., and Georges, S., Direct methane solid oxide fuel cell working by gradual internal steam reforming: Analysis of operation, 331–337, Copyright 2009, with permission from Elsevier.)

simplifying the fuel cell system. However, most SOFC systems still require pre-reforming or internal reforming processes to avoid carbon deposition and mitigate fuel cell degradation.

2.5.2.2 External Reforming

Although internal reforming of hydrocarbon fuels can improve fuel cell efficiency while providing additional cooling for the cell, using external reformers in stationary power plants to avoid carbon deposition in anode chambers as well as large temperature gradients in SOFC stacks has been proposed [59,60]. For external reforming, there are four methods developed [3,61]:

1. Steam reforming (SR). The reactions are as follows:

$$CH_4 + H_2O \rightarrow CO + 3H_2 \tag{2.15}$$

$$CO + H_2O \rightarrow CO_2 + H_2O \tag{2.16}$$

2. Partial-oxidation reforming (POX):

$$CH_4 + \frac{1}{2}O_2 \rightarrow CO + 2H_2 \tag{2.17}$$

3. Auto-thermal reforming (ATR):

$$2CH_4 + O_2 + CO_2 \rightarrow 3CO + 3H_2 + H_2O \tag{2.18}$$

$$2CH_4 + \frac{1}{2}O_2 + H_2O \rightarrow 5H_2 + 2CO \tag{2.19}$$

4. Thermal decomposition (TDC):

$$CH_4 \rightarrow C + 2H_2 \tag{2.20}$$

Steam–methane reforming consumes steam to produce hydrogen from methane, ethanol, propane, gasoline, and diesel fuel streams. In steam–methane reforming, hydrocarbon fuels react with steam to produce H_2, CO, and CO_2. Steam reforming reactions are endothermic and require external heat sources. Carbon monoxide can further react with steam to produce more hydrogen in a water–gas shifting process. Catalytic partial oxidation of hydrocarbons is another attractive alternative for SOFC systems since the reaction is weakly exothermic, which will largely simplify the thermal integration of the reforming process and SOFCs. The ATR process can be regarded as a combination of SR and POX and is more flexible than steam reforming considering start-up times and load change characteristics.

Membrane reactors have attracted considerable attention for their compactness and ability to implement separation and reaction at the same time [62]. There are two types of commonly used membrane reactors for fuel processing: oxygen ion transport membrane (OTM) reformers and hydrogen transport membrane (HTM) reformers. OTM reformers avoid the direct mixing of fuels and O_2 that is common in partial oxidation reactors. This isolation of fuel from air eliminates nitrogen-associated pollution emissions. HTM reformers, as highly compact reactors, also possess advantages such as rich hydrogen product gases and high fuel conversion ratios.

2.5.3 CO$_2$ CAPTURE IN SOFC-BASED POWER GENERATION SYSTEMS

In SOFCs, fuel is oxidized by oxygen ions without the dilution of CO_2 through air. Therefore, fuel consumption in SOFCs can be considered to be an oxy-combustion process. The electrolyte can be regarded as a oxygen permeable membrane. The driving force of the oxygen separation is the chemical potential difference between both sides of the oxygen permeable membrane. Therefore, oxygen separation in this membrane reactor is performed without additional energy and can effectively reduce efficiency penalties and costs due to CO_2 capture. Thus, SOFCs are one of the most promising candidates for carbon-free power generation in the near future.

Based on the current technology status of SOFCs, there are many different technical choices for CO_2 capture in SOFC-based power systems. For instance, pre-SOFC and post-SOFC CO_2 capture, usually including physical absorption [63–65], cryogenics [66,67], and membranes [68,69], are often considered. After the pretreatment of syngas fuels using a shifting reactor followed by the physical adsorption of H_2

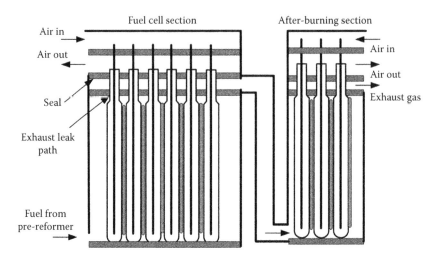

FIGURE 2.17 CO_2 separating SOFC concept. (Reprinted from *J. Power Sour.*, 106, Haines, M.R., Heidug, W.K., Li, K.J., and Moore, J.B., Progress with the development of a CO_2 capturing solid oxide fuel cell, 377–380, Copyright 2002, with permission from Elsevier.)

using an H_2 selective membrane, highly purified H_2 can be obtained. The higher H_2 content fuel can effectively improve SOFC performance, efficiency, and fuel utilization ratio, while decreasing the risk of carbon deposition. By feeding simple fuels, such as H_2, to SOFC stacks, internal reforming and shifting processes are no longer needed, which decreases fuel cell degradation with more uniformly distributed temperatures and gas compositions.

To better demonstrate the special advantages of SOFCs in carbon capture, many researchers have focused on the complete oxidation of residue fuels in anode flue gases without CO_2 dilution, noted as post-SOFC CO_2 capture technology. Figure 2.17 shows one SOFC system designed by Siemens–Westinghouse Power Generation [70]. By adding additional fuel cell tubes, referred to as a "after burning" section, the fuel utilization ratio can be increased from 85% to around 98%.

Inui et al. [71] proposed an SOFC/GT system incorporating the function of CO_2 separation where exhaust gases from the SOFC stack are burnt by adding pure O_2. The pure oxygen used results in significant additional costs and energy consumption however. Many researchers have tried to use oxygen ion transport membranes (OTM) as the afterburner in SOFC systems to supply oxygen across the membrane to implement H_2 and CO combustion in SOFC exhaust gases.

Jansen et al. [72,73] proposed an SOFC system integrated with a water–gas shift membrane reactor (WGSMR) to convert CO in the exhaust gas into CO_2. The working principle of this hybrid membrane reactor system is shown in Figure 2.18. They introduced three types of H_2 selective membranes, including a microporous membrane, a palladium membrane, and a proton-conducting membrane. The working temperature of the proton-conducting membrane was relatively high,

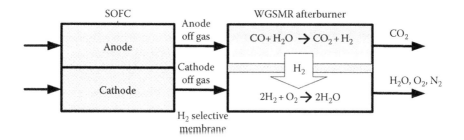

FIGURE 2.18 Working principles of the WGSMR afterburner. (Reprinted from *Energy*, 29, Dijkstra, J.W. and Jansen, D., Novel concepts for CO_2 capture, 1249–1257, Copyright 2004, with permission from Elsevier.)

which is beneficial to integration with SOFCs. This is, however, still at the early stages of development. Compared with OTM, WGSMR possesses special advantages for CO_2 capture in SOFC systems. For example, mass flux across the membrane moves toward the cathode depleted gases and this can be fed to the gas turbine to produce more power due to the increase in mass flow. The H_2 flux can also be maintained at a reasonable value without additional power consumption, which will improve system efficiency.

From the discussions earlier, it is clear that SOFC technologies are promising for power generation combined with CO_2 capture. There are also many potential technologies for fuel cell system integration, such as additional SOFC afterburners, OTM afterburners, and WGSMR afterburners. These potential technologies require further studies to meet commercialization requirements.

2.6 SUMMARY

SOFC technology is a feasible power technology for clean electricity with significant efforts being put forth by various companies toward commercialization.

In regard to technological developments, particularly on SOFC electrode and electrolyte materials, although Ni-YSZ-based anode materials have demonstrated stability during fuel cell operations, more advanced materials are still required to lower the operation temperatures of SOFC systems. This is because operating SOFC systems at intermediate temperatures (500°C–700°C) will significantly prolong the lifetime of a fuel cell stack and reduce the costs of stack components. Currently, advanced cathodes and electrolytes are being intensively studied to reduce ohmic resistances and speed up oxygen reduction reaction kinetics at lower temperatures. Fuel conversion mechanisms at the anode also require further research. Although precommercialized SOFC products have appeared in the market, the performance, reliability/durability, and lifetime need to be significantly improved. In particular, their costs need to be greatly reduced to meet commercialization requirements. BoP instruments of SOFC systems, such as desulfurize reactors, water purifiers for SMR reactions, as well as afterburners for system start-up and emission control, also need to be improved and their costs reduced.

REFERENCES

1. Williams M.C., Strakey J.P., Surdoval W.A., Wilson L.C. 2006. Solid oxide fuel cell technology development in the U.S. *Solid State Ionics* 177: 2039–2044.
2. Bove R., Ubertini S. 2008. *Modeling Solid Oxide Fuel Cells: Methods, Procedures and Techniques.* Springer Science & Business Media, Berlin/Heidelberg, Germany.
3. EG&G Technical Services, Inc. 2004. *Fuel Cell Handbook*, 7th edn. Contract No. DE-AM26-99FT40575, U.S. Department of Energy, Office of Fossil Energy, National Energy Technology Laboratory, WV.
4. Minh N.Q. 2004. Solid oxide fuel cell technology—Features and applications. *Solid State Ionics* 174: 271–277.
5. Shi Y. et al. 2007. Modeling of an anode-supported Ni–YSZ| Ni–ScSZ| ScSZ| LSM–ScSZ multiple layers SOFC cell: Part I. experiments, model development and validation. *Journal of Power Sources* 172: 235–245.
6. Shi Y. et al. 2007. Modeling of an anode-supported Ni-YSZ| Ni-ScSZ| ScSZ| LSM-ScSZ multiple layers SOFC cell: Part II. Simulations and discussion. *Journal of Power Sources* 172: 246–252.
7. Singhal S.C., Kendall K. 2003. *High-Temperature Solid Oxide Fuel Cells: Fundamentals, Design and Applications.* Elsevier, Amsterdam, Netherlands.
8. Goodenough J.B. 2000. Ceramic technology: Oxide-ion conductors by design. *Nature* 404: 821–823.
9. Tsipis E.V., Kharton V.V. 2008. Electrode materials and reaction mechanisms in solid oxide fuel cells: A brief review. *Journal of Solid State Electrochemistry* 12: 1039–1060.
10. Lefebvre-Joud F., Gauthier G., Mougin J. 2009. Current status of proton-conducting solid oxide fuel cells development. *Journal of Applied Electrochemistry* 39: 535–543.
11. Fleig J. 2003. Solid oxide fuel cell cathodes: Polarization mechanisms and modeling of the electrochemical performance. *Annual Review of Materials Research* 33: 361–382.
12. Zhu W.Z., Deevi S.C. 2003. A review on the status of anode materials for solid oxide fuel cells. *Materials Science and Engineering: A* 362: 228–239.
13. Brandon N.P., Skinner S., Steele B.C.H. 2003. Recent advances in materials for fuel cells. *Annual Review of Materials Research* 33: 183–213.
14. Gross M.D., Vohs J.M., Gorte R.J. 2007. Recent progress in SOFC anodes for direct utilization of hydrocarbons. *Journal of Materials Chemistry* 17: 3071.
15. Bieberle A. 2000. *The Electrochemistry of Solid Oxide Fuel Cell Anodes: Experiments, Modeling, and Simulation.* Swiss Federal Institute of Technolgoy, Zurich, Switzerland.
16. Bieberle A., Gauckler L.J. 2002. State-space modeling of the anodic SOFC system Ni, H_2–H_2O|YSZ. *Solid State Ionics* 146: 23–41.
17. Mogensen M.B., Lindegaard T. 1993. The kinetics of hydrogen oxidation on a Ni-YSZ SOFC electrode at 1000°C. In *Symposium on Solid Oxide Fuel Cells.* Pennington, NJ.
18. de Boer B. 1998. *SOFC Anode: Hydrogen Oxidation at Porous Nickel and Nickel/Yttria-Stabilised Zirconia Cermet Electrodes.* Universiteit Twente, Enschede, the Netherlands.
19. Goodwin D. 2005. A pattern anode model with detailed chemistry. In *Ninth International Symposium on Solid Oxide Fuel Cells (SOFC-IX),* Quebec, May.
20. Zhu H., Kee R.J., Janardhanan V.M., Deutschmann O., Goodwin D.G. 2005. Modeling elementary heterogeneous chemistry and electrochemistry in solid-oxide fuel cells. *Journal of the Electrochemical Society* 152: A2427–A2440.
21. Mizusaki J. et al. 1994. Preparation of nickel pattern electrodes on YSZ and their electrochemical properties in H_2-H_2O atmospheres. *Journal of the Electrochemical Society* 141: 2129–2134.

22. Eguchi K., Kojo H., Takeguchi T., Kikuchi R., Sasaki K. 2002. Fuel flexibility in power generation by solid oxide fuel cells. *Solid State Ionics* 152: 411–416.

23. Sukeshini A.M., Habibzadeh B., Becker B.P., Stoltz C.A., Eichhorn B.W., Jackson G.S. 2006. Electrochemical oxidation of H_2, CO, and CO/H_2 mixtures on patterned Ni anodes on YSZ electrolytes. *Journal of the Electrochemical Society* 153: A705–A715.

24. Matsuzaki Y., Yasuda I. 2000. Electrochemical oxidation of H_2 and CO in a H_2-H_2O-CO-CO_2 system at the interface of a Ni-YSZ cermet electrode and YSZ electrolyte. *Journal of the Electrochemical Society* 147: 1630–1635.

25. Eisen I.H., Hengas S.N. 1971. Overpotential behavior of stabilized zirconia solid electrolyte fuel cells. *Journal of the Electrochemical Society* 118: 1890–1900.

26. Mizusaki J. et al. 1992. Kinetics of the electrode reaction at the CO-CO_2, porous Pt/ stabilized zirconia interface. *Solid State Ionics* 53: 126–134.

27. Jiang Y., Virkar A.V. 2003. Fuel composition and diluent effect on gas transport and performance of anode-supported SOFCs. *Journal of the Electrochemical Society* 150: A942–A951.

28. Costa-Nunes O., Gorte R.J., Vohs J.M. 2005. Comparison of the performance of Cu–CeO_2–YSZ and Ni–YSZ composite SOFC anodes with H_2, CO, and syngas. *Journal of Power Sources* 141: 241–249.

29. Habibzadeh B. 2007. *Understand CO Oxidation in SOFC's Using Nickel Patterned Anode*. University of Maryland, College Park, MD.

30. Murray E.P., Tsai T., Barnett S.A. 1999. A direct-methane fuel cell with a ceria-based anode. *Nature* 400: 649–651.

31. Putna E.S., Stubenrauch J., Vohs J.M., Gerte R.J. 1995. Ceria-based anodes for the direct oxidation of methane in solid oxide fuel cells. *Langmuir* 11(12): 4832–4837.

32. Park S., Craciun R., Vohs J.M., Gorte R.J. 1999. Direct oxidation of hydrocarbons in a solid oxide fuel cell: I. Methane oxidation. *Journal of the Electrochemical Society* 146: 3603–3605.

33. Anderson J.R., Boudart M. 2012. *Catalysis: Science and Technology*. Springer Science & Business Media, Berlin/Heidelberg, Germany.

34. Kazakov A., Wang H., Frenklach M. 1995. Detailed modeling of soot formation in laminar premixed ethylene flames at a pressure of 10 bar. *Combustion and Flame* 100: 111–120.

35. Appel J., Bockhorn H., Frenklach M. 2000. Kinetic modeling of soot formation with detailed chemistry and physics: Laminar premixed flames of C_2 hydrocarbons. *Combustion and Flame* 121: 122–136.

36. Walters K.M., Dean A.M., Zhu H., Kee R.J. 2003. Homogeneous kinetics and equilibrium predictions of coking propensity in the anode channels of direct oxidation solid-oxide fuel cells using dry natural gas. *Journal of Power Sources* 123: 182–189.

37. Sheng C.Y., Dean A.M. 2004. Importance of gas-phase kinetics within the anode channel of a solid-oxide fuel cell. *The Journal of Physical Chemistry A* 108: 3772–3783.

38. Besenbacher F. et al. 1998. Design of a surface alloy catalyst for steam reforming. *Science* 279: 1913–1915.

39. Rostrup-Nielsen J.R., Alstrup I. 1999. Innovation and science in the process industry: Steam reforming and hydrogenolysis. *Catalysis Today* 53: 311–316.

40. Triantafyllopoulos N.C., Neophytides S.G. 2006. Dissociative adsorption of CH_4 on NiAu/YSZ: The nature of adsorbed carbonaceous species and the inhibition of graphitic C formation. *Journal of Catalysis* 239: 187–199.

41. Costamagna P., Costa P., Antonucci V. 1998. Micro-modelling of solid oxide fuel cell electrodes. *Electrochimica Acta* 43: 375–394.

42. Costamagna P., Costa P., Arato E. 1998. Some more considerations on the optimization of cermet solid oxide fuel cell electrodes. *Electrochimica Acta* 43: 967–972.

43. Chan S.H., Xia Z.T. 2001. Anode micro model of solid oxide fuel cell. *Journal of the Electrochemical Society* 148: A388–A394.

44. Virkar A.V., Chen J., Tanner C.W., Kim J. 2000. The role of electrode microstructure on activation and concentration polarizations in solid oxide fuel cells. *Solid State Ionics* 131: 189–198.

45. Huang K. 2004. Gas-diffusion process in a tubular cathode substrate of an SOFC I theoretical analysis of gas-diffusion process under cylindrical coordinate system. *Journal of the Electrochemical Society* 151: A716–A719.

46. Recknagle K.P., Williford R.E., Chick L.A., Rector D.R., Khaleel M.A. 2003. Three-dimensional thermo-fluid electrochemical modeling of planar SOFC stacks. *Journal of Power Sources* 113: 109–114.

47. Burt A.C., Celik I.B., Gemmen R.S., Smirnov A.V. 2004. A numerical study of cell-to-cell variations in a SOFC stack. *Journal of Power Sources* 126: 76–87.

48. Lin B., Shi Y., Ni M., Cai N. 2015. Numerical investigation on impacts on fuel velocity distribution nonuniformity among solid oxide fuel cell unit channels. *International Journal of Hydrogen Energy* 40: 3035–3047.

49. Lin B., Shi Y., Cai N. 2017. Numerical simulation of cell-to-cell performance variation within a syngas-fuelled planar solid oxide fuel cell stack. *Applied Thermal Engineering* 114: 653–662.

50. Yi Y., Smith T.P., Brouwer J., Rao A.D., Samuelsen G.S. 2003. Simulation of a 220 kW hybrid SOFC gas turbine system and data comparison. In *Eighth International Symposium on Solid Oxide Fuel Cells (SOFC-VIII)*, Paris, April.

51. Williams M.C., Strakey J.P., Singhal S.C. 2004. US distributed generation fuel cell program. *Journal of Power Sources* 131: 79–85.

52. Chaney J., Tharp R., Wolf W., Fuller A., Hartvigson J. 1999. *Fuel Cell/Micro Turbine Combined Cycle.* McDermott Technology, Inc., Alliance, OH, DOE Contract No. DE-AC26-98FT40454.

53. Lobachyov K., Richter H.J. 1996. Combined cycle gas turbine power plant with coal gasification and solid oxide fuel cell. *Journal of Energy Resources Technology* 118: 285–292.

54. Samms S.R., Savinell R.F. 2002. Kinetics of methanol-steam reformation in an internal reforming fuel cell. *Journal of Power Sources* 112: 13–29.

55. Nagata S., Momma A., Kato T., Kasuga Y. 2001. Numerical analysis of output characteristics of tubular SOFC with internal reformer. *Journal of Power Sources* 101: 60–71.

56. Brus G., Szmyd J.S. 2008. Numerical modelling of radiative heat transfer in an internal indirect reforming-type SOFC. *Journal of Power Sources* 181: 8–16.

57. Vernoux P., Guindet J., Kleitz M. 1998. Gradual internal methane reforming in intermediate-temperature solid-oxide fuel cells. *Journal of the Electrochemical Society* 145: 3487–3492.

58. Klein J., Hénault M., Roux C., Bultel Y., Georges S. 2009. Direct methane solid oxide fuel cell working by gradual internal steam reforming: Analysis of operation. *Journal of Power Sources* 193: 331–337.

59. Peters R., Riensche E., Cremer P. 2000. Pre-reforming of natural gas in solid oxide fuel-cell systems. *Journal of Power Sources* 86: 432–441.

60. Peters R., Dahl R., Klüttgen U., Palm C., Stolten D. 2002. Internal reforming of methane in solid oxide fuel cell systems. *Journal of Power Sources* 106: 238–244.

61. Chan S.H., Ho H.K., Tian Y. 2003. Multi-level modeling of SOFC–gas turbine hybrid system. *International Journal of Hydrogen Energy* 28: 889–900.

62. Qi A., Peppley B., Karan K. 2007. Integrated fuel processors for fuel cell application: A review. *Fuel Processing Technology* 88: 3–22.

63. Al-Ghawas H.A., Hagewiesche D.P., Ruiz-Ibanez G., Sandall O.C. 1989. Physico-chemical properties important for carbon dioxide absorption in aqueous methyldietha-nolamine. *Journal of Chemical and Engineering Data* 34: 385–391.

64. Bai H.L., Yeh A.C. 1997. Removal of CO_2 greenhouse gas by ammonia scrubbing. *Industrial & Engineering Chemistry Research* 36: 2490–2493.

65. Huang Y., Rezvani S., McIlveen-Wright D., Minchener A., Hewitt N. 2008. Techno-economic study of CO_2 capture and storage in coal fired oxygen fed entrained flow IGCC power plants. *Fuel Processing Technology* 89: 916–925.

66. Clodic D, Younes M. 2002. A new method for CO_2 capture: Frosting CO_2 at atmospheric pressure. Presented at *Sixth International Conference on Greenhouse Gas control Technologies GHGT6*, Kyoto, Japan.

67. Hart A., Gnanendran N. 2009. Cryogenic CO_2 capture in natural gas. *Energy Procedia* 1: 697–706.

68. Bredesen R., Jordal K., Bolland O. 2004. High-temperature membranes in power gener-ation with CO_2 capture. *Chemical Engineering and Processing: Process Intensification* 43: 1129–1158.

69. Ho M.T., Allinson G.W., Wiley D.E. 2008. Reducing the cost of CO_2 capture from flue gases using membrane technology. *Industrial & Engineering Chemistry Research* 47: 1562–1568.

70. Haines M.R., Heidug W.K., Li K.J., Moore J.B. 2002. Progress with the development of a CO_2 capturing solid oxide fuel cell. *Journal of Power Sources* 106: 377–380.

71. Inui Y., Matsumae T., Koga H., Nishiura K. 2005. High performance SOFC/GT com-bined power generation system with CO_2 recovery by oxygen combustion method. *Energy Conversion and Management* 46: 1837–1847.

72. Jansen D., Dijkstra J.W. May 2003. CO_2 capture in SOFC-GT systems. Presented at *Second Annual Conference on Carbon Sequestration*, Alexandria, VA.

73. Dijkstra J.W., Jansen D. 2004. Novel concepts for CO_2 capture. *Energy* 29: 1249–1257.

3 Solid Oxide Electrolysis Cells

3.1 INTRODUCTION

Electrolysis is a process in which electrical power is converted into chemical energy. For instance, electricity can split H_2O into oxygen and hydrogen in an electrolysis cell. Hydrogen, a clean energy carrier, is promising for applications in next-generation power generation and transportation [1–3]. Presently, there are three types of electrolysis technologies being widely applied or studied: alkaline electrolysis cells (AECs), proton exchange membrane electrolysis cells (PEMECs), and solid oxide electrolysis cells (SOECs). AECs and PEMECs usually operate at relatively low temperatures (<200°C), whereas SOECs operate at high temperatures (500°C–1000°C).

High-temperature electrolysis through SOECs was first explored in the 1960s by NASA to provide submarines and spacecraft with oxygen to support life and propulsion [3–5]. Further studies of hydrogen production using SOECs were carried out in the 1980s [6]. In general, high temperatures lead to lower polarization losses and replace roughly 30% of required electricity inputs [3]. Therefore, SOECs can perform at high power-to-hydrogen efficiencies. The fast electrode reaction kinetics at high temperatures also allow for the use of low-cost catalysts such as nickel. Furthermore, the considerable heat required to support high temperature operations of SOECs can be attained from integration with renewable sources such as photothermal energy or geothermal energy as well as from nuclear energy to allow for efficient generation of clean fuels [6,7].

From another perspective, global warming as well as depleting fossil fuels have raised growing interest in CO_2 capture, utilization, and storage (CCUS). SOECs using ionic-conducting electrolytes make it possible to effectively electrolyze H_2O and CO_2, providing an alternative pathway to convert CO_2 and H_2O into syngas ($CO + H_2$) and pure oxygen [3,8–11]. The syngas produced can then be widely used to produce methanol, gasoline, diesel, and other hydrocarbon fuels through the Fischer–Tropsch (F-T) synthesis process. The integration of SOECs and F-T synthesis is one of the most promising and viable pathways to convert H_2O and CO_2 into fuels [3,12]. Nuclear energy, being a clean, low-carbon-producing power source with negligible carbon dioxide emissions, can drive SOECs [3,13]. Renewable power sources such as wind and solar power can also drive SOECs to produce fuels. On the other hand, the intermittence, specific site, and fluctuation of these renewable power sources could impact the grid, leading to difficulties in its applications [14]. This intermittence of renewable energy sources can be shifted and stabilized using reversible solid oxide cell (RSOC) technologies where electricity is stored in the form of chemical energy via SOECs and then utilized to generate electricity via solid oxide fuel cells (SOFCs). The viability of RSOCs provides an alternative pathway for the storage of seasonal power [3,15–17].

The advantages of SOECs have attracted increasing interests for large-scale fuel production and CCUS. The current development of SOEC technologies is however still in early stages of commercialization. A series of issues, optimizations, and improvements need to be resolved and improved upon, including reaction mechanisms, performance enhancements, production orientation conversions, reliability and durability for long-term operations, system integration, costs, and more [3]. This chapter will provide a comprehensive review of the fundamentals of SOECs and related studies from microlevel elementary reaction mechanisms to macrolevel system integration and optimizations.

3.2 FUNDAMENTALS OF THE SOLID OXIDE ELECTROLYSIS CELL

SOEC is the inverse process of SOFC. Similar to SOFCs, the structure of the positive electrode–electrolyte–negative electrode assembly (PEN) is an essential part of SOECs in which electrochemical reactions take place. PEN is also referred to in other articles as MEA (Membrane Electrode Assembly). In the schematic diagram of a SOEC as shown in Figure 3.1, steam and carbon dioxide react in a porous negative electrode (NE) with incoming electrons from an external circuit, generating hydrogen, carbon monoxide, and oxygen ions. Oxygen ions then travel across a densely structured electrolyte to a porous structured positive electrode (PE). In this porous positive electrode, the oxygen ions react to generate gaseous oxygen and electrons [3,18]. Aside from electrochemical reactions, heterogeneous surface reactions containing reversible water–gas shift (WGS) reactions, methanation reactions as well as carbon deposition reactions can take place at the negative electrode as well.

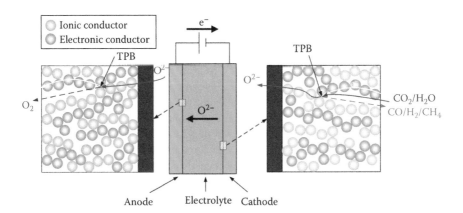

FIGURE 3.1 Schematic diagram of solid oxide electrolysis cells. (Reprinted from *J. Power Sources*, 243, Li, W., Shi, Y., Luo, Y., and Cai, N., Elementary reaction modeling of CO_2/H_2O co-electrolysis cell considering effects of cathode thickness, 118–130, Copyright 2013, with permission from Elsevier.)

3.2.1 BASIC STRUCTURES AND WORKING PRINCIPLES

As shown in Figure 3.1, a PEN possesses a structure where an ionic-conducting electrolyte layer is sandwiched between a porous negative electrode and a positive electrode. The dense electrolyte provides pathways for oxygen ion conduction as well as a barrier for effective separation between the positive electrode chamber and the negative electrode chamber [19]. H_2O or/and CO_2 is fed into the negative electrode channel while air or oxygen flows into the positive electrode channel. Porous electrodes generally consist of a mixture of ionic-conductive and electronic-conductive particles with the purpose of improving the density of the triple-phase boundary (TPB). TPB, the interface between the electronic conductor phase, ionic conductor phase, and gaseous phase, is the reaction zone that is commonly considered to be the electrochemical reactive site. H_2O or/and CO_2 diffuses from the negative electrode chamber to TPB sites and combines with electrons. The half-reaction at the negative electrode can be expressed as

$$H_2O + 2e^- \rightarrow H_2 + O^{2-} \tag{3.1}$$

$$CO_2 + 2e^- \rightarrow CO + O^{2-} \tag{3.2}$$

Oxygen ions are released at the negative electrode and conducted through the electrolyte to the TPB of the negative electrode, losing electrons and producing oxygen. The half-reaction at the positive electrode can be expressed as

$$O^{2-} \rightarrow \frac{1}{2}O_2 + 2e^- \tag{3.3}$$

The overall reactions of SOECs are expressed as

$$H_2O \rightarrow H_2 + \frac{1}{2}O_2 \tag{3.4}$$

$$CO_2 \rightarrow CO + \frac{1}{2}O_2 \tag{3.5}$$

In addition, heterogeneous surface catalytic processes are significant for negative electrodes. Nickel metal is a typical metal inside negative electrodes serving as an electronic conductor and a catalyst. Nickel catalysts can significantly accelerate the reaction rates of reversible WGS and methanation reactions, therefore, the negative electrodes of SOECs are crucial not only for electrochemical reactions, but also important for the promotion of heterogeneous reactions. This is the basis of the direct conversion of syngas/methane in the co-electrolysis process of H_2O and CO_2. Moreover, carbon can be produced through the Boudouard reaction, leading to a series of issues such as pore blocking, catalyzer deactivation, polarization loss growth, nickel loss, and mechanical failure through fiber growth [20–24]. These processes can be described by the following reactions:

$$H_2 + CO_2 \underset{Ni}{\rightleftharpoons} H_2O + CO \quad \Delta H = 41 \text{ kJ/mol} \tag{3.6}$$

$$3H_2 + CO \xrightarrow{Ni} CH_4 + H_2O \quad \Delta H = -206 \text{ kJ/mol} \tag{3.7}$$

$$2CO \rightleftharpoons CO_2 + C \quad \Delta H = -183 \text{ kJ/mol} \tag{3.8}$$

3.2.2 THERMODYNAMICS

Thermodynamics is a governing criteria for both chemical and electrochemical reactions. Whether a reaction occurs spontaneously or not in a certain condition can be determined and the minimum energy required to sustain electrochemical reactions can also be theoretically predicted based on thermodynamics [25,26].

Under open circuit condition, the cell voltage, referred to as the open circuit voltage (OCV), denotes the minimum electricity demand. Here, OCV depends on the reversible Nernst potential, E_N, if the reactant crossover the electrolyte can be ignored. In thermodynamics, the change in Gibbs free energy, ΔG, in SOECs represents the minimum electrical power required to power a certain reaction. Therefore, theoretical OCV, also referred to as Nernst potential E_N, is proportional to ΔG based on the following equation:

$$\Delta G = n_e F E_N \tag{3.9}$$

Furthermore, because ΔG is related to temperature and gas composition, theoretical OCV E_N can be further expressed as [27]

$$E_N = \frac{\Delta G^0}{n_e F} - \frac{\Delta S}{n_e F}(T - 298.15) + \frac{RT}{n_e F} \ln \frac{\prod p_{products}^{\upsilon_i}}{\prod p_{reactants}^{\upsilon_i}} \tag{3.10}$$

In Equation 3.10, ΔG^0 denotes ΔG in standard conditions, ΔS denotes the entropy change, n_e denotes the transferred electron number, F denotes Faraday's constant (96,485°C/mol), R denotes the universal gas constant, T denotes the temperature (K), p denotes the partial pressure of a certain species, and υ_i denotes the stoichiometric coefficients in the overall reaction equation. Taking H_2O electrolysis as an example, E_N is expressed as

$$E_N = \frac{\Delta G^0}{n_e F} - \frac{\Delta S}{n_e F}(T - 298.15) + \frac{RT}{n_e F} \ln \frac{p_{H_2} p_{O_2}^{1/2}}{p_{H_2O}} \tag{3.11}$$

When the negative electrode side is fed with $H_2O:H_2$ (0.5:0.5) and the positive electrode side with air, E_N is equal to 1.263 V at 100°C, while decreasing to 1.016 V at 800°C. Total energy demanded by a reaction is calculated based on equation below.

$$\Delta H = \Delta G + T\Delta S \qquad (3.12)$$

For SOECs, the electrolysis process is endothermic. Therefore, SOECs operating at a constant temperature demand both electrical power and thermal energy. In Equation 3.12, ΔG denotes the electrical energy demand and $T\Delta S$ denotes the heat demand. Here, ΔH, equal to the sum of ΔG and $T\Delta S$, represents the total energy demand including electricity and heat during the electrolysis process. Figure 3.2 shows the energy demands of both H_2O and CO_2 electrolysis in temperatures ranging from 0°C to 1000°C [3]. It can be found out that the electric power required for splitting H_2O and CO_2 decreases as the operation temperature grows higher. But heat needed for the reaction increases when the temperature rises (at temperatures higher than 100°C). The calculation based on thermodynamics indicates that high-temperature electrolysis by SOECs can effectively use heat energy (such as the waste heat from power plants, chemical industries, etc.) to replace electrical energy. Taking steam electrolysis for example, electrolyzing 1 mol of H_2O requires 225.2 kJ of electricity at 100°C and 188.7 kJ of electricity at 800°C. The temperature elevation here provided an electricity savings of 16%, which is significant in terms of saving electricity and improving efficiency for the conversion of power to gas.

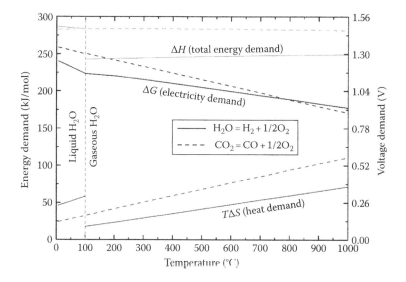

FIGURE 3.2 Energy demands for H_2O and CO_2 electrolysis in temperatures ranging from 0°C to 1000°C. (Shi, Y., Luo, Y., Li, W., Ni, M., and Cai, N.: High temperature electrolysis for hydrogen or syngas production from nuclear or renewable energy. Yan, J. (ed.), *Handbook of Clean Energy Systems*. New York. 2015. Copyright Wiley-VCH Verlang GmbH &m Co. KGaA. Reproduced with permission.)

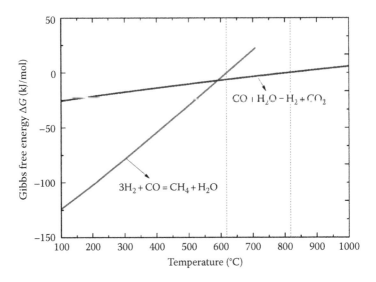

FIGURE 3.3 Gibbs free energy of reversible WGS and methanation reactions in temperatures ranging from 100°C to 1000°C. (Shi, Y., Luo, Y., Li, W., Ni, M., and Cai, N.: High temperature electrolysis for hydrogen or syngas production from nuclear or renewable energy. Yan, J. (ed.), *Handbook of Clean Energy Systems*. New York. 2015. Copyright Wiley-VCH Verlang GmbH &m Co. KGaA. Reproduced with permission.)

Thermodynamics also determines reaction directions and equilibriums of heterogeneous catalytic reactions. Based on thermodynamics, a reaction is spontaneous when the Gibbs free energy ΔG is less than 0. Both reversible WGS and methanation reactions in SOECs can be predicted by their thermodynamics. Figure 3.3 shows the ΔG values of WGS and methanation reactions in temperatures ranging from 100°C to 1000°C. The data indicate that reversible WGS reactions are favored to occur in the forward direction at temperatures below 820°C and methanation reactions are favored to occur in the forward direction at temperatures below 620°C. Thus, hydrogen and carbon monoxide production can be promoted by reversed WGS (RWGS) in temperatures below 820°C, while methane production is expected to only occur below 620°C.

3.2.3 REACTION KINETICS OF SOECs

Thermodynamics indicate the limits of electrolysis in SOECs, whereas reaction kinetics dominates the electrolysis rates of SOECs. Although reactions in SOECs can theoretically occur when cell voltages are higher than the reversible potentials E_N based on thermodynamics, these reactions could be slow due to limited kinetics. The practical performance of SOECs can be affected by complicated couplings of chemistry/electrochemistry and mass transport processes. Cell performances depend on three types of polarization losses: ohmic polarization caused by ionic/electronic charge transfer resistances, activation polarization caused by the irreversibility of electrochemistry, and concentration polarization caused by gas concentration differences within the porous electrodes. Consequently, an actual operating voltage larger than E_N must be applied to overcome these irreversible losses to obtain desired reaction rates. The driving potential for overcoming these irreversible losses is called overpotential η. The lower the overpotential, the better the SOEC can perform when operating at a certain reaction rate.

The three losses mentioned earlier suggest that the overall overpotential η can be divided into three parts: active overpotential η_{act}, ohmic overpotential η_{ohm}, and concentration overpotential η_{conc}. Considering these three potentials, the operational cell voltage V_{cell} can be expressed as

$$V_{cell} = E_N + \eta_{act} + \eta_{ohm} + \eta_{conc} \tag{3.13}$$

For electrode kinetics, the Butler–Volmer (B-V) equation is widely applied in the description of electrochemical reaction kinetics [3,28]:

$$
\begin{aligned}
i_{NE} &= i_{0,NE} \left[\frac{c_{P,NE}}{c_{P,NE}^0} \exp\left(\frac{\alpha_{NE} n_e F \eta_{act,NE}}{RT} \right) - \frac{c_{R,NE}}{c_{R,NE}^0} \exp\left(-\frac{(1-\alpha_{NE}) n_e F \eta_{act,NE}}{RT} \right) \right] \\
i_{PE} &= i_{0,PE} \left[\frac{c_{R,PE}}{c_{R,PE}^0} \exp\left(\frac{\alpha_{PE} n_e F \eta_{act,PE}}{RT} \right) - \frac{c_{P,PE}}{c_{P,PE}^0} \exp\left(-\frac{(1-\alpha_{PE}) n_e F \eta_{act,PE}}{RT} \right) \right]
\end{aligned}
\tag{3.14}
$$

where

i_{NE} and i_{PE} denote the current densities passing through the negative electrode and positive electrode

$i_{0,NE}$ and $i_{0,PE}$ denote the exchange current densities of the negative and positive electrode

c_R and c_P denote the concentrations of the reactant and product

c_R^0 and c_P^0 denote the reference concentrations of the reactant and product

α_{NE} and α_{PE} denote the symmetric parameters of the negative and positive electrode

R and T in Equation 3.14 are the universal gas constant and the temperature

In Equation 3.14, the first term describes the electrochemical reaction rate of the negative direction and the second term of the positive one. The exchange current density is a key parameter to evaluate cell performance, and is determined by operating conditions such as temperature, gas composition, electrode materials, as well as electrode microstructures related to TPB densities [29,30].

Ohmic overpotential η_{ohm} expressed by Equation 3.15 mainly powers ion transfer through the electrolyte. The related ohmic resistance R is determined by ionic conductivity of the electrolyte and electrolyte thickness. In practical operation of an SOEC stack, however, the ohmic resistance of the interconnector and contact layer can also influence ohmic overpotential.

$$\eta_{ohm} = iR \tag{3.15}$$

To reduce the ohmic losses, making the electrolyte thinner is a potential development trend. Furthermore, temperatures also strongly influence the ionic conductivity of electrolytes. When SOECs are operate at a given temperature, the ohmic overpotential is almost linear to the current input.

Concentration overpotential η_{conc} is mainly caused by the the lower concentration of reactants in the porous electrode than that in the flow channel of electrode chamber. Taking H_2O electrolysis as an example, the negative electrode–electrolyte interface is the reactive region for the electrochemical reactions, with lower H_2O concentration and more abundant H_2 than on the outer surface of the negative electrode. As for the positive electrode–electrolyte interface, O_2 is more concentrated than that on the outer surface of the positive electrode. Consequently, the concentration difference between the electrode–electrolyte interface and the electrode outer surface increase the partial pressure–related terms in the Nernst equation (Equation 3.11), increasing the value of E_N. The increase of E_N is considered as the concentration overpotential. Gas concentrations can also affect the reaction kinetics of the B-V equation (Equation 3.14) and are considered as the activation overpotential η_{act}. The effects of concentration overpotentials become increasingly significant with increasing operation currents.

3.3 REACTION MECHANISMS IN NICKEL-PATTERNED ELECTRODES

Nickel-patterned electrodes are ultrathin pure nickel electrodes possessing two-dimensional, regular-shaped structures. Compared to porous three-dimensional electrodes, Ni-patterned electrodes have quantifiable TPB lengths and Ni surface areas, allowing the simple characterization of surface topographies as well as the exclusion of bulk gas transports in porous structures [20,31–35]. Thus, Ni-patterned electrodes are employed to understand reaction mechanisms and to acquire intrinsic kinetic parameters for both SOFCs and SOECs [36,37]. Reaction mechanisms, degradation, and impurity effects have been intentionally studied for SOFCs using Ni-patterned electrodes [31,38–41]. Few studies on Ni-patterned electrodes have focused on mechanisms of SOECs however. The authors tested Ni-patterned electrodes with a stripe width of 100 μm and thickness of 800 nm fabricated on monocrystal 13 mol% YSZ electrolytes. Figure 3.4 shows the geometric dimensioning of the tested single crystal cell [42]. The Pt electrode tested was fabricated through mesh printing on the other side of the single crystal cell. The length of the TPB was calculated to be 364.3 ± 0.6 mm, the Ni area 19.07 ± 0.09 mm^2, and the TPB density 19.10 mm/mm^2.

Stability of the Ni-patterned electrode at high temperatures was demonstrated during testing, as shown in Figure 3.5 [42]. Hole formations on the Ni surface can separate the Ni-patterned electrode into Ni islands and disrupt the regular Ni pattern. It was found that reducing power-on time (≤ 8 h), thickening Ni-patterned electrode (≥ 800 nm), lowering partial pressure of $CO + CO_2$ (≤ 0.5 atm), and reducing temperature ($\leq 700°C$) and overpotential (≤ 0.3 V) can all effectively prevent hole formation on the Ni surface [42]. Figure 3.5 shows that the width of the Ni stripe did not shrink and the edges (TPB) remained binded with the single crystal YSZ although the size of the Ni grains increased and the Ni surface roughened after the electrochemical tests. This demonstrates that Ni-patterned electrodes remain stable at high-temperature operations with unchanged designed sizes of TPB and Ni surfaces.

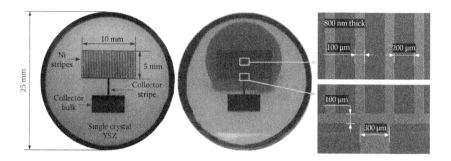

FIGURE 3.4 Photos of the Ni-patterned electrode before and after mesh printing the Pt electrode and microscopic SEM images before testing [20,42]. (Reprinted from *J. Power Sources*, 276, Li, W., Shi, Y., Luo, Y., Wang, Y., and Cai, N., Carbon deposition on patterned nickel/yttria stabilized zirconia electrodes for solid oxide fuel cell/solid oxide electrolysis cell modes, 26–31, Copyright 2015, with permission from Elsevier; Reprinted from *Int. J. Hydrogen Energy*, 41, Li, W., Shi, Y., Luo, Y., Wang, Y., and Cai, N., Carbon monoxide/carbon dioxide electrochemical conversion on patterned nickel electrodes operating in fuel cell and electrolysis cell modes, 3762–3773, Copyright 2016, with permission from Elsevier.)

FIGURE 3.5 Microscopic SEM images of Ni patterned electrode after testing. (Reprinted from *Int. J. Hydrogen Energy*, 41, Li, W., Shi, Y., Luo, Y., Wang, Y., and Cai, N., Carbon monoxide/carbon dioxide electrochemical conversion on patterned nickel electrodes operating in fuel cell and electrolysis cell modes, 3762–3773, Copyright 2016, with permission from Elsevier.)

3.3.1   CARBON DEPOSITION MECHANISM

3.3.1.1   Distribution of Carbon Deposition

Our previous article [20] provides a detailed investigation into carbon depositions. In this article, various Ni-patterned electrodes were tested at 750°C with a negative electrode inlet CO_2/CO molar ratio of 1.0, and applied voltages of OCV (0.942 V), OCV ± 0.2 V, and OCV ± 0.7 V. After testing, the Ni surfaces were characterized by energy-dispersive X-ray spectroscopy (EDS) to estimate elemental atomic percentages. Carbon generation from the Boudouard reaction (Equation 3.8) was found led to similar carbon atomic percentages in the middle and edge of the nickel stripe after being operated under OCV conditions. In SOFC mode, the carbon atomic percentages at low overpotentials were significantly higher than that at OCV, indicating that low current densities promote Boudouard reaction kinetics [22].

The carbon atomic percentages reduced in SOFC mode running at high overpotentials however, suggesting that high current densities decelerate carbon generation. Similar conclusions were found in porous-Ni electrode SOFCs fueled with methane at 700°C and 800°C [43,44]. The carbon atomic percentages in the center of the Ni stripe (Ni surface) were larger than that on the edge (TPB), indicating that C could take part in electrochemical reactions and could be consumed in SOFC mode. The electrochemical oxidation of deposited carbon into CO at the TPB was speculated due to the easier formation of C–O bonds than C=O bonds [42].

In SOEC mode, opposite phenomena were observed. When the overpotential increased from 0.2 to 0.7 V, the C at% in the center of and on the edge of the Ni stripe both increased but the C at% near the TPB was higher, suggesting that electricity input can dramatically improve carbon deposition through an electrochemical route. Furthermore, C=O bond energy is over twice that of C–O bond energy, providing a possible way of carbon deposition at the TPB under SOEC mode via the CO electrochemical reduction into carbon.

To summarize the phenomena described earlier and in the previous article [45], the highly possible mechanisms of carbon deposition are demonstrated in Figure 3.6 and Table 3.1, which simplifies the reaction mechanisms proposed by Deutschmann's group [46]. Gaseous CO/CO_2 first adsorbs on the Ni surface (Ni)

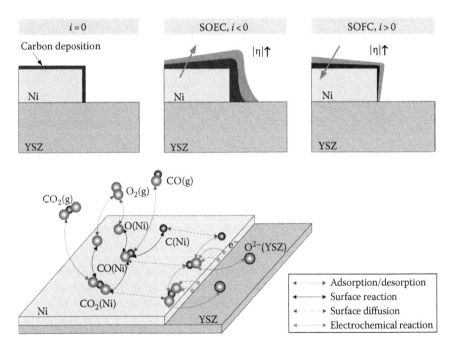

FIGURE 3.6 Schematic diagram of carbon deposition in different modes and the related elementary reactions. (Reprinted from *J. Power Sources*, 276, Li, W., Shi, Y., Luo, Y., Wang, Y., and Cai, N., Carbon deposition on patterned nickel/yttria stabilized zirconia electrodes for solid oxide fuel cell/solid oxide electrolysis cell modes, 26–31, Copyright 2015, with permission from Elsevier.)

TABLE 3.1
Elementary Reactions in Figure 3.6

Adsorption and desorption on Ni surface

$$CO_2(g) + (Ni) \leftrightarrow CO_2(Ni) \tag{3.16}$$

$$CO(g) + (Ni) \leftrightarrow CO(Ni) \tag{3.17}$$

$$O(Ni) + O(Ni) \leftrightarrow O_2(g) + (Ni) + (Ni) \tag{3.18}$$

Surface reaction on Ni surface

$$CO_2(Ni) + (Ni) \leftrightarrow CO(Ni) + O(Ni) \tag{3.19}$$

$$CO(Ni) + (Ni) \leftrightarrow C(Ni) + O(Ni) \tag{3.20}$$

Charge transfer reaction

$$O(Ni) + (YSZ) + 2e^- \underset{SOFC}{\overset{SOEC}{\rightleftharpoons}} (Ni) + O^{2-}(YSZ) \tag{3.21}$$

$$CO(Ni) + (YSZ) + 2e^- \underset{SOFC}{\overset{SOEC}{\rightleftharpoons}} C(Ni) + O^{2-}(YSZ) \tag{3.22}$$

Source: Reprinted from *J. Power Sources*, 276, Li, W., Shi, Y., Luo, Y., Wang, Y., and Cai, N., Carbon deposition on patterned nickel/yttria stabilized zirconia electrodes for solid oxide fuel cell/solid oxide electrolysis cell modes, 26–31, Copyright 2015, with permission from Elsevier.

as shown by Equations 3.16 and 3.17. Then Equations 3.19 and 3.20 show that the adsorbed $CO(Ni)$, $CO_2(Ni)$, and the free surface active sites (Ni) can react on the surface of the Ni stripe and generate adsorbed $O(Ni)$ and $C(Ni)$. Equation 3.20 shows that the adsorbed $O(Ni)$ and gaseous $O_2(Ni)$ can be converted into each other through adsorption and desorption processes. When a current is applied, two charge transfer reactions (Equations 3.21 and 3.22) would take place, involving the adsorbed $CO_2(Ni)$, $CO(Ni)$, and $C(Ni)$. Deposited carbon can be generated by SOEC and consumed by SOFC in the electrochemical reaction, as shown in Equation 3.22.

3.3.1.2 Structure Features of Deposited Carbon

In our tests, elemental carbon distribution was detected and its mechanism was proposed based on EDS analysis [20]. The structural features of the deposited carbon were obtained using Raman spectroscopy. Here, only one peak was detected within the range of 1350 cm^{-1} and 1590 cm^{-1}, which correspond the G band of the Raman spectra. The G band is used to detect C–C bond in the graphite structure. The Raman spectra of carbon deposited under OCV mode indicated that regular crystal graphitic carbon structures were the dominant structures in the carbon deposition, and the amorphous carbon was minor when no electricity was introduced. In SOEC mode, the deposited carbon near the TPB (on the edge of Ni stripe) possessed a higher graphitic carbon content than that on the Ni surface (in the center), particularly at high polarization voltages. However, graphitic carbon is found more prone to be consumed under SOFC mode by electrochemical oxidation. Disordered carbon and defect carbon were not found on the Ni surface or near the TPB, which is different from CH$_4$-fueled SOFCs [47] but consistent with our previous study on CO-fueled SOFCs [22]. This suggests a different carbon deposition mechanism of CO-fueled SOFCs from CH$_4$-fueled SOFCs. Under CO/CO$_2$ atmosphere, graphitic carbon is the primary type of carbon participating in the electrochemical reactions of SOFCs/SOECs.

3.3.2 ELECTROCHEMICAL CONVERSION MECHANISMS OF CO/CO$_2$

In our previous study [42], the effects of partial pressure, p_{CO} or p_{CO_2}, and temperature on electrochemical performances were investigated to acquire intrinsic kinetic parameters of SOEC reactions. The possible electrochemical conversion mechanisms and rate-limiting steps were speculated.

3.3.2.1 Electrochemical Reaction Kinetics

The total overpotential (η_{tot}) of an Ni-YSZ-Pt SOEC can be expressed as Equation 3.23:

$$\eta_{tot} = \eta_{Ni} + \eta_{ohm} + \eta_{Pt} \qquad (3.23)$$

where

η_{Ni} is the overpotential of the Ni-patterned electrode
η_{ohm} is the overpotential of the YSZ
η_{Pt} is the overpotential of the Pt electrode

To separate Ni-patterned electrode overpotentials from total overpotential, symmetric Pt electrode cells printed on both sides of a single crystal YSZ electrolyte were tested in air. The ohmic overpotential and Pt electrode overpotential here were almost identical in SOFC and SOEC modes, accounting for only 0.6%–2.9% and 0.9%–3.8% of the total overpotential, respectively [42]. Therefore, the electrochemical performances of the total cells can approximately represent that of Ni-patterned electrodes.

The B-V equation (Equation 3.14) describes the relationship between the overpotential η and the current density i. In the B-V equation, the exchange current density i_0 relies on the temperature and partial pressure of the species participating in the reaction. In the negative electrode side, i_0 can be expressed in the Arrhenius form as [40,42]

$$i_0 = \gamma \left(p_{CO} \right)^a \left(p_{CO_2} \right)^b \exp\left(-\frac{E_{act}}{RT} \right) \qquad (3.24)$$

where

γ is the pre-exponential factor
a is the dependency of p_{CO}
b is the dependency of p_{CO_2}
E_{act} is the activation energy of CO/CO$_2$ electrochemical conversion

In the B-V equation, the relationship between the overpotential and current density is almost linear at low overpotential ranges. Hence, the corresponding polarization resistance is inversely proportional to i_0. According to Equation 3.24, the key parameters a, b, and E_{act} can be calculated from measured experimental data to evaluate the effects of T, p_{CO}, and p_{CO_2} using Equations 3.25 through 3.27:

$$E_{act} = R \frac{d \ln(R_{pol}/T)}{d(1/T)} \bigg|_{\partial OV, p_{CO}, p_{CO_2}} \tag{3.25}$$

$$a = -\frac{\partial \ln(R_{pol})}{\partial \ln(p_{CO})} \bigg|_{\eta^*, T} \tag{3.26}$$

$$b = -\frac{\partial \ln(R_{pol})}{\partial \ln(p_{CO_2})} \bigg|_{\eta^*, T} \tag{3.27}$$

where R_{pol} is the polarization resistance of the Ni-patterned electrode. The applicability of the B-V equation is based on the premise that the charge transfer reaction is the rate-limiting step [42]. If the B-V equation is unable to correlate with experimental data, it suggests that the electrochemical performance could be co-limited by other physical/chemical processes, such as surface diffusion, adsorption or desorption, and surface chemistry. But, to some extent, Equations 3.25 through 3.27 are also helpful in analyzing electrochemical processes.

3.3.2.2 Effects of Applied Voltage

The effects of applied voltage can be evaluated by EIS and R_{pol} as shown in Figure 3.7 [42]. Polarization voltages help the SOEC to overcome activation barriers, which is in accordance with results in SOFC modes. R_{pol} can be evaluated by the difference between the two points where the EIS curve intersects on the real axis. The EIS curves indicate that the performance of SOECs is dominated by activation polarization as proofed by the increasing R_{pol} with rising overpotentials.

Polarization resistances will first increase significantly with increased applied voltages in SOEC mode however. The EIS curve indicates that the shape of the low-frequency zone can also be affected by diffusion resistance. Therefore, surface diffusion is also important in the CO_2 electrolysis process. R_{pol} begins to decrease when the Ni-patterned electrode overpotential η_{Ni} is further increased beyond -0.25 V in SOEC mode. This phenomenon is speculated to be caused by the changing of surface diffusion coefficients (D_{sf}). D_{sf} relies on surface features such as crystal forms and surface coverage [42,48]. The D_{sf} of surface species is small, with large surface diffusion resistances at low overpotentials in SOEC mode. At high overpotentials, the D_{sf} of surface species increases with decreasing surface coverage, and if the overpotential is large enough to consume most of the surface species, it leads to a smaller surface diffusion resistance. Consequently, surface diffusion is speculated to be one of the co-limiting steps at the negative electrode at low overpotentials in SOEC mode. Increasing TPB density and decreasing Ni stripe width can avoid this.

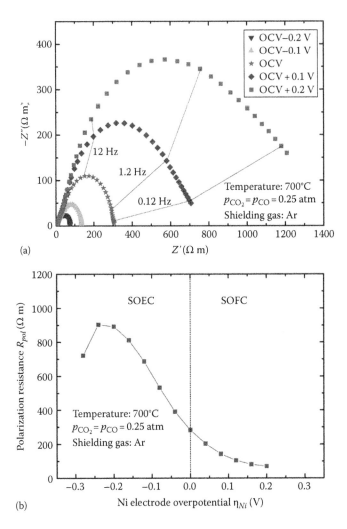

FIGURE 3.7 Effects of applied voltage: (a) EIS curve; (b) R_{pol}. (Reprinted from *Int. J. Hydrogen Energy*, 41, Li, W., Shi, Y., Luo, Y., Wang, Y., and Cai, N., Carbon monoxide/carbon dioxide electrochemical conversion on patterned nickel electrodes operating in fuel cell and electrolysis cell modes, 3762–3773, Copyright 2016, with permission from Elsevier.)

3.3.2.3 Effects of Temperature

Figure 3.8 demonstrates the polarization curves, polarization resistances, and EIS curves at different temperatures [42]. The polarization resistance of Ni electrodes in both SOFC and SOEC modes dramatically increases with decreasing temperatures. The shapes of the EIS curves indicate that although activation resistances dominate cell performance, diffusion resistances also play a role. The surface diffusion resistance, reflected qualitatively by $R_{pol,max} - R_{pol,OCV}$, increases dramatically in SOEC mode with decreasing temperature. According to Equation 3.25, the activation energy can be acquired through a linearly fitting curve of $\ln R_{pol,OCV} - 1/T$. The fitting value of E_{act} at OCV is 1.77 eV.

3.3.2.4 Effects of CO Partial Pressure

Figure 3.9 demonstrates the polarization curves, polarization resistances, and EIS curves at different CO partial pressures [42]. The increase of CO partial pressure has a significant negative effect on the polarization resistances of Ni electrodes in both SOFC and SOEC modes. The current ratios of SOFC to SOEC modes increase with increasing CO partial pressures. Figure 3.9b further indicates that the approximate surface diffusion resistance $R_{pol,max} - R_{pol,OCV}$ decreases with increasing p_{CO}, implying that CO(Ni) could be a key surface species for surface diffusion.

The linear fitting of $\ln R_{pol,OCV} - p_{CO}$ data was conducted to obtain the parameter a according to Equation 3.26. At 700°C and $p_{CO_2} = 0.25$ atm, the value of a at OCV

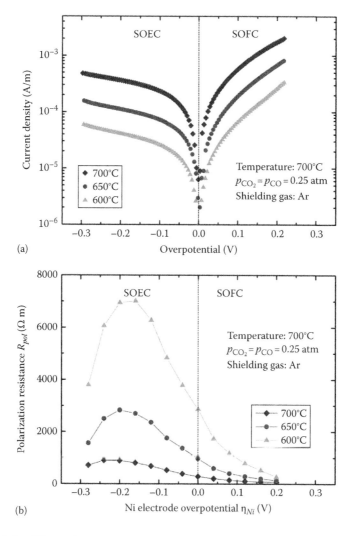

(a)

(b)

FIGURE 3.8 Effects of temperature: (a) polarization curve; (b) R_{pol}.   (*Continued*)

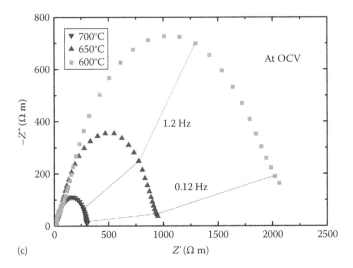

(c)

FIGURE 3.8 (*Continued*) Effects of temperature: (c) EIS curve. (Reprinted from *Int. J. Hydrogen Energy*, 41, Li, W., Shi, Y., Luo, Y., Wang, Y., and Cai, N., Carbon monoxide/carbon dioxide electrochemical conversion on patterned nickel electrodes operating in fuel cell and electrolysis cell modes, 3762–3773, Copyright 2016, with permission from Elsevier.)

was 0.657, demonstrating a positive correlation between R_{pol} and p_{CO}. The value of a increases with increasing η_{Ni} from −0.24 to 0.16 V, suggesting that the effects of p_{CO} on the polarization resistance of SOFCs are more dramatic than that on SOECs. This is because CO can serve as the main reactant in SOFC modes and is one of the main products in SOEC modes.

3.3.2.5   Effects of CO$_2$ Partial Pressure

Figure 3.10 shows the polarization curves, polarization resistances, and EIS curves at different CO$_2$ partial pressures [42]. CO$_2$ partial pressure has a limited effect on the polarization resistance of Ni electrodes in both SOFC and SOEC modes, owing to the slow CO$_2$ adsorption on the surface of highly purified Ni at elevated temperatures [42,49,50]. This phenomenon is different from that in porous Ni-YSZ electrodes, where p_{CO_2} is closely related with electrochemical performances. The metal oxide in porous electrode can provide lattice oxygen and defects, which make the surface prone to adsorption, not to mention the enhancing effect on adsorption of the electric field [42,51]. The porous Ni-YSZ electrode containing metal oxides could absorb more CO$_2$ and possesses a TPB length density of 10^{12} m/m^3, which is three orders of magnitude higher than that of the Ni-patterned electrode [52]). Therefore, performance of porous Ni-YSZ electrodes are not limited by CO$_2$ adsorption and have a close relationship with p_{CO_2}. The linear fitting of $\ln R_{pol,OCV} - p_{CO_2}$ data was conducted to obtain the parameter b according to Equation 3.27. At 700°C and $p_{CO} = 0.25$ atm, the value of b at OCV was 0.011. The positive value of b in SOEC mode reveals the slight positive effects of p_{CO_2} on the performance of Ni-patterned electrodes, while the negative value in SOFC mode reveals a slight negative effects.

3.3.2.6 Reaction Mechanism Speculations for CO_2 Electrolysis

Based on the earlier experimental data and analysis, several possible rate-determining steps are considered and the following simplified reaction mechanisms are proposed [42]:

1. CO_2 adsorbs near the TPB on the Ni surface:

$$CO_2(g) + (Ni) \rightarrow CO_2(Ni) \tag{3.28}$$

2. CO_2 decomposes on the Ni surface:

$$CO_2(Ni) + (Ni) \leftrightarrow CO(Ni) + O(Ni) \tag{3.29}$$

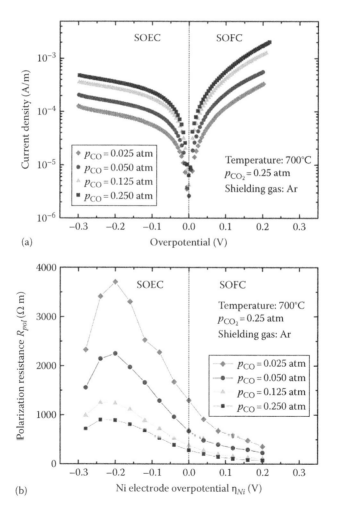

(a)

(b)

FIGURE 3.9 Effects of CO partial pressure: (a) polarization curve; (b) R_{pol}. *(Continued)*

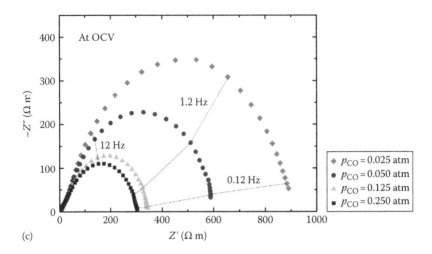

FIGURE 3.9 (*Continued*)   Effects of CO partial pressure: (c) EIS curve. (Reprinted from *Int. J. Hydrogen Energy*, 41, Li, W., Shi, Y., Luo, Y., Wang, Y., and Cai, N., Carbon monoxide/carbon dioxide electrochemical conversion on patterned nickel electrodes operating in fuel cell and electrolysis cell modes, 3762–3773, Copyright 2016, with permission from Elsevier.)

3. O(Ni) diffuses to the TPB on the Ni surface (a surface diffusion process)
4. Charge transfer reaction at the TPB:

$$O(Ni) + (YSZ) + 2e^- \rightarrow (Ni) + O^{2-}(YSZ) \tag{3.30}$$

5. CO desorbs on the Ni surface:

$$CO(Ni) \rightarrow CO(g) + (Ni) \tag{3.31}$$

3.4   HETEROGENEOUS CHEMISTRY AND ELECTROCHEMISTRY OF SOEC POROUS ELECTRODES

In practical applications, porous structures are the most common structures for SOEC electrodes to provide adequate mechanical strength and gas diffusion paths. A common material system for SOEC NEs is Ni-YSZ, which is also used in SOFCs due to its high conductivity, catalytic activity, economic viability, good mechanical strength, and chemical stability. Nickel is a widely accepted catalyst in promoting the reaction rates of reversible WGS and methanation under CO_2/H_2O atmosphere. There are numerous papers on carbon deposition in Ni-YSZ electrodes. Therefore, the heterogeneous chemistry and electrochemistry of porous electrodes are worth studying to evaluate the mechanisms of the reactions in the Ni-YSZ electrode.

3.4.1 BASIC PERFORMANCE

The performance of the porous Ni-YSZ electrode button cells tested by our group is shown in Figure 3.11 [53]. The button cells were made up of one 680 μm negative electrode support layer (Ni-YSZ), one 15 μm negative electrode active layer (nickel–Scandia-stabilized zirconia [Ni-ScSZ]), one 20 μm electrolyte layer (ScSZ), and one 15 μm positive electrode layer (LSM-ScSZ). Figure 3.11a shows the significant positive dependency of temperature on cell performances. The addition of H_2 also led to a higher OCV in the case at 730°C. Performance of the cell is significantly improved when hydrogen is introduced into the reaction system. Figure 3.11b shows the effects of various gas compositions. The three gas compositions denote three

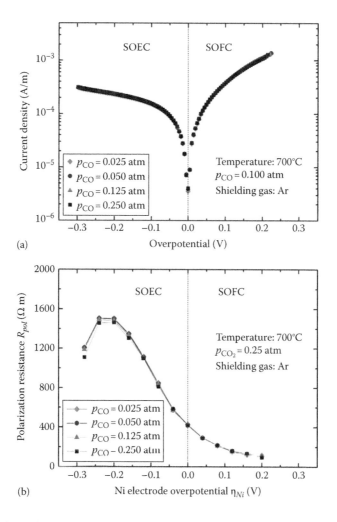

(a)

(b)

FIGURE 3.10 Effects of CO_2 partial pressure: (a) polarization curve; (b) R_{pol}. (*Continued*)

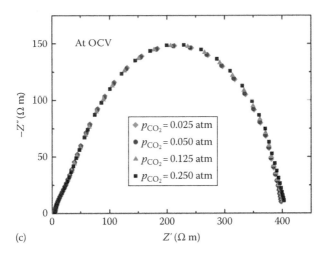

(c)

FIGURE 3.10 (Continued)   Effects of CO_2 partial pressure: (c) EIS curve. (Reprinted from *Int. J. Hydrogen Energy*, 41, Li, W., Shi, Y., Luo, Y., Wang, Y., and Cai, N., Carbon monoxide/carbon dioxide electrochemical conversion on patterned nickel electrodes operating in fuel cell and electrolysis cell modes, 3762–3773, Copyright 2016, with permission from Elsevier.)

electrolysis modes: H_2O electrolysis, CO_2 electrolysis, and H_2O/CO_2 co-electrolysis. The electrochemical performance of H_2O electrolysis was the best, with H_2O/CO_2 co-electrolysis lying between the performance of H_2O and CO_2 electrolysis. This suggests that not only can CO be produced both from CO_2 electrolysis and RWGS in the co-electrolysis mode. This conclusion is in accordance with the results from Risø National Laboratory [14] but is different from the results from Idaho National Laboratory [54] and Mahidol University [55]. Idaho National Laboratory [54] and Mahidol University [55] observed that the performance of the co-electrolysis mode was almost identical to that of the H_2O electrolysis; hence, CO in their results was almost exclusively produced from RWGS and little from CO_2 electrolysis. The crucial difference found between these two conclusions is the NE thickness. The former adopted a NE-supported SOEC with hundreds of micrometers in NE thickness, whereas the latter adopted an electrolyte-supported SOEC with only tens of micrometers in thickness. The difference will be explained clearly in Section 3.4.4.3.

3.4.2 Analysis of Methane Production Pathways

According to thermodynamics, the standard Gibbs free energy of methanation (ΔG_{ME}) becomes negative only at temperatures below 620°C. In our experiment, methane was detected in the co-electrolysis mode [53]. It has been observed in the literature [53] that when hydrogen is introduced, methane content is increased by more than eightfold as operating voltages increase from OCV to 2 V. Therefore, CH_4 is speculated to be generated from a pathway related to electricity. When Ru is further impregnated into the porous NE, proven to reduce carbon deposition, reversed

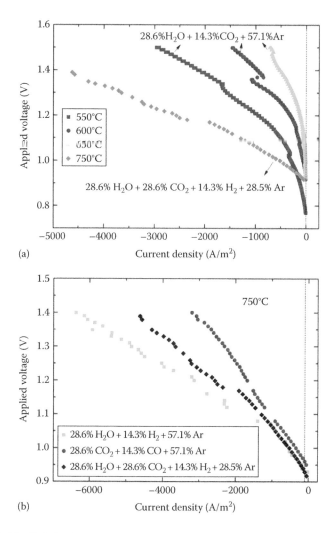

FIGURE 3.11 Polarization curves: (a) different temperatures; (b) different gas compositions. (Reprinted from *Int. J. Hydrogen Energy,* 38, Li, W., Wang, H., Shi, Y., and Cai, N., Performance and methane production characteristics of H_2O–CO_2 co-electrolysis in solid oxide electrolysis cells, 11104–11109, Copyright 2013, with permission from Elsevier.)

methanation (steam reforming of methane) [53,56] and WGS [53,57] are promoted. Ru impregnation leads to less CO content and more H_2 content in the product due to the promotion of WGS with methane being greatly reduced from 0.286% to 0.077% at 2 V [53]. The decrease in methane production is thought to be caused by not only the promotion of reversed methanation, but also the suppression of carbon deposition. Therefore, the reaction of $C(s) + 2H_2 \rightarrow CH_4$ can be an important reaction pathway for methane generation. In summary, the complete reaction pathways in relation to elemental carbon and its corresponding reactions (R1–R7) are proposed in

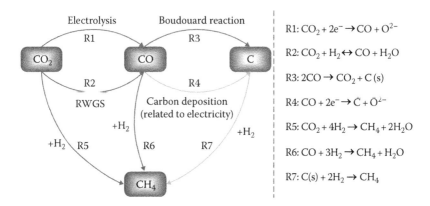

FIGURE 3.12   Reaction pathways related to elemental carbon in the H_2O/CO_2 co-electrolysis process of SOECs. (From Li, W. et al., *Int. J. Hydrogen Energy*, 11104, 2013.)

Figure 3.12. R3, R4, and R7 are related to carbon deposition while R5–R7 are related to methane production. By adding hydrogen, R5 and R6 can be greatly improved to produce more CH_4.

3.4.3   Elementary Reaction Models and PEN Models

Based on the proposed reaction mechanisms, elementary reaction models can be built to predict detailed processes of catalytic reactions and electrochemistry reactions. The elementary reaction processes may contain several steps, such as gas adsorption/desorption, surface chemical reaction, surface and bulk mass transfer as well as charge transfer reactions [3]. It is crucial to understand chemistry/electrochemistry mechanisms and kinetic parameters in order to clarify their effects on negative electrode microstructures and dependencies on materials as well as working conditions. Furthermore, elementary reaction studies are also needed to determine the dominating rate-determining steps in the mechanisms. Hence, reaction mechanism models describing chemical/electrochemical reaction kinetics, conversion pathways of surface species, and combined effects of related parameters are needed [3,58]. Operating conditions and materials can also greatly influence rate-determining steps, causing different rate-determining steps at different operating conditions [3,59].

Mechanistic PEN models usually incorporate mass transfer, charge transfer as well as reaction mechanisms. Mechanistic PEN models use conservation equations to describe the coupled reaction and transport processes, and could reveal the distributions of species in the PEN. Because of this, the impacts of geometries, electrode microstructures, and material properties of cells can be predicted in this type of PEN models.

An elementary reaction model for the co-electrolysis of CO_2/H_2O in SOECs coupled with a PEN model has been developed in our group [35,45,58,60,61] and provides a useful tool for the identification of reaction and transport mechanisms. This model is helpful in explaining measured data and building connections between cell design and performance optimization. This detailed heterogeneous reaction mechanism is based on our previous works as well as literature [46], as shown in Table 3.2.

TABLE 3.2
Heterogeneous Reaction Mechanisms on Ni-Catalyst Surfaces

Number of Reactions	Elementary Reaction	A (cm, mol, s)[a]	n[a]	E (kJ/mol)[a]
	Adsorption and desorption			
1[f]	$H_2(g) + (Ni) + (Ni) \rightarrow H(Ni) + H(Ni)$	1.000×10^{-02}[b]	0.0	0.00
1[r]	$H(Ni) + H(Ni) \rightarrow H_2(g) + (Ni) + (Ni)$	$2.545 \times 10^{+19}$	0.0	81.21
2[f]	$O_2(g) + (Ni) + (Ni) \rightarrow O(Ni) + O(Ni)$	1.000×10^{-02}[b]	0.0	0.00
2[r]	$O(Ni) + O(Ni) \rightarrow O_2(g) + (Ni) + (Ni)$	$4.283 \times 10^{+23}$	0.0	474.95
3[f]	$H_2O(g) + (Ni) \rightarrow H_2O(Ni)$	0.100×10^{-00}[b]	0.0	0.00
3[r]	$H_2O(Ni) \rightarrow H_2O(g) + (Ni)$	$3.732 \times 10^{+12}$	0.0	60.79
4[f]	$CO_2(g) + (Ni) \rightarrow CO_2(Ni)$	1.000×10^{-05}[b]	0.0	0.00
4[r]	$CO_2(Ni) \rightarrow CO_2(g) + (Ni)$	$6.447 \times 10^{+07}$	0.0	25.98
5[f]	$CO(g) + (Ni) \rightarrow CO(Ni)$	5.000×10^{-02}[b]	0.0	0.00
5[r]	$CO(Ni) \rightarrow CO(g) + (Ni)$	$3.563 \times 10^{+11}$	0.0	111.27
		$\theta_{CO(Ni)}$		−50.00[c]
	Surface reaction			
6[f]	$H(Ni) + O(Ni) \rightarrow OH(Ni) + (Ni)$	$5.000 \times 10^{+22}$	0.0	97.90
6[r]	$OH(Ni) + (Ni) \rightarrow H(Ni) + O(Ni)$	$1.781 \times 10^{+21}$	0.0	36.09
7[f]	$H(Ni) + OH(Ni) \rightarrow H_2O(Ni) + (Ni)$	$3.000 \times 10^{+20}$	0.0	42.70
7[r]	$H_2O(Ni) + (Ni) \rightarrow H(Ni) + OH(Ni)$	$2.271 \times 10^{+21}$	0.0	91.76
8[f]	$OH(Ni) + OH(Ni) \rightarrow H_2O(Ni) + O(Ni)$	$3.000 \times 10^{+21}$	0.0	100.00
8[r]	$H_2O(Ni) + O(Ni) \rightarrow OH(Ni) + OH(Ni)$	$6.373 \times 10^{+23}$	0.0	210.86
9[f]	$CO(Ni) + O(Ni) \rightarrow CO_2(Ni) + (Ni)$	$2.000 \times 10^{+19}$	0.0	123.60
		$\theta_{CO(Ni)}$		−50.00[c]
9[r]	$CO_2(Ni) + (Ni) \rightarrow CO(Ni) + O(Ni)$	$4.653 \times 10^{+23}$	−1.0	89.32
	Charge transfer reaction			
10[f]	$O(Ni) + (YSZ) + 2e^- \xrightarrow{SOEC} (Ni) + O^{2-}(YSZ)$	$\dfrac{i_0}{FS_{TPB}c^0_{O(Ni)}c^0_{(YSZ)}}$	0	$2(1-\alpha)F\eta_{NE}$
10[r]	$(Ni) + O^{2-}(YSZ) \xrightarrow{SOFC} O(Ni) + (YSZ) + 2e^-$	$\dfrac{i_0}{FS_{TPB}c^0_{(Ni)}c^0_{O^{2-}(YSZ)}}$	0	$2\alpha F\eta_{NE}$

Sources: Reprinted from *J. Power Sources*, 162, Janardhanan, V.M. and Deutschmann O., CFD analysis of a solid oxide fuel cell with internal reforming: Coupled interactions of transport, heterogeneous catalysis and electrochemical processes, 1192–1202, Copyright 2006, with permission from Elsevier; Reprinted from *J. Power Sources*, 243, Li, W., Shi, Y., Luo, Y., and Cai, N., Elementary reaction modeling of CO_2/H_2O co-electrolysis cell considering effects of cathode thickness, 118–130, Copyright 2013, with permission from Elsevier.

[a] Rate constant written in the Arrhenius form: $k - AT^n\exp(-E/RT)$.

[b] Sticking coefficient.

[c] Coverage-dependent activation energy.

f, forward reaction; r, reverse reaction.

3.4.4 EFFECTS OF KEY PARAMETERS ON THE NEGATIVE ELECTRODE PROCESS

To better understand the electrochemical process, the effects of key parameters on current density and species distribution should be identified, including the reaction rate, applied voltage, NE thickness, electrode structure, ionic conductivity, temperature, gas composition, and so on

3.4.4.1 Sensitivity Analysis of Heterogeneous Elementary Reaction Rates

To get a deeper understanding of SOEC processes, the key elementary reactions has to be identified. Therefore, a sensitivity analysis to evaluate the influence of reversible reactions 1–9 in Table 3.2 on the cell performances is carried out, as shown in Figure 3.13 [58]. The relative significance among these adsorption/desorption reactions is CO_2 adsorption (No. 4^f) ≈ CO desorption (No. 5^r) < H_2O adsorption (No. 3^f) < H_2 desorption (No. 1^r). In addition, the following surface reactions are also significant: $H_2O(Ni) + (Ni) \rightarrow H(Ni) + OH(Ni)(R7^r)$, $OH(Ni) + (Ni) \rightarrow H(Ni) + O(Ni)$ $(R6^r)$, and $CO_2(Ni) + (Ni) \rightarrow CO(Ni) + O(Ni)(R9^r)$. Focusing on the conversion of CO_2 into CO $(R4^f \rightarrow R9^r \rightarrow R5^r)$ and H_2O into H_2 $(R3^f \rightarrow R7^r \rightarrow R6^r \rightarrow R1^r)$, it can be seen that hydrogen production from steam is faster than carbon monoxide production from carbon dioxide. Hence, the key element O(Ni) is more easily obtained from steam decomposition than from carbon dioxide decomposition, implying that H_2O electrolysis performs better than CO_2 electrolysis.

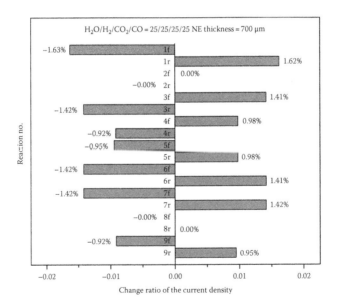

FIGURE 3.13 Sensitivity of the heterogeneous elementary reaction rate to the current density. (Reprinted from *J. Power Sources*, 243, Li, W., Shi, Y., Luo, Y., Cai, N., Elementary reaction modeling of CO_2/H_2O co-electrolysis cell considering effects of cathode thickness, 118–130, Copyright 2013, with permission from Elsevier.)

3.4.4.2   Effects of Applied Voltage

Figure 3.14 shows the effects of applied voltage on the distributions of current density and surface species concentration. The current density and surface species concentration at OCV are variable within a 200 μm region from the negative electrode–electrolyte interface, especially in the negative electrode active layer with the better ionic conductivity and higher TPB density. Free site (Ni), adsorbed H(Ni), and adsorbed CO(Ni) are the three primary surface species on the Ni surface of negative electrodes. Voltage increases the surface coverage of adsorbed CO(Ni) and adsorbed H(Ni) but decreases the surface coverage of adsorbed O(Ni) OH(Ni), CO_2(Ni), H_2O(Ni), and free site (Ni).

3.4.4.3   Effects of Negative Electrode Thickness

Figure 3.15 shows the polarization curves of three electrolysis modes with negative electrode thicknesses of 30, 60, 100, and 700 μm in atmospheres with an oxidant/reductant ratio of 1.0 [58]. As the thickness of the negative electrode decreases, the performance of H_2O/CO_2 co-electrolysis becomes increasingly similar to that of H_2O electrolysis. The curves of these two electrolysis modes almost overlap at a negative electrode thickness of 30 μm. If the negative electrode is too thin (30 μm) however, the electrochemical performance will be limited by the amount of TPB active sites. Thin electrodes are beneficial to gas and ionic diffusions however. Therefore, there are optimal values of negative electrode thicknesses for optimal performance. In our simulation, this value is 60 μm. This value is not universal however, because optimal thicknesses can be affected by electrode microstructures with different pore diameters, porosities, or tortuosity [58].

To further clarify how negative electrode thicknesses can influence heterogeneous chemistry and electrochemistry processes, reaction rates of the dominating heterogeneous elementary reaction and charge transfer reaction at 1.4 V in three electrolysis modes for negative electrode thicknesses of 700, 100, 60, and 30 μm are displayed in Figure 3.16. It can be seen that the reaction rate of the charge transfer reaction is two orders of magnitude lower than that of the heterogeneous elementary reaction, indicating that the charge transfer reaction is the crucial rate-limiting step of the electrolysis process. The electrochemical reaction rate near the interface of the negative electrode and the electrolyte of steam/carbon dioxide co-electrolysis is between that of steam electrolysis and carbon dioxide electrolysis modes, and becomes closer to that of H_2O electrolysis with decreasing negative electrode thicknesses. Free site (Ni) and adsorbed O(Ni) are two important surface species related to the charge transfer reaction and their concentration distributions are demonstrated in Figure 3.17. The surface coverage of free site (Ni) and adsorbed O(Ni) in steam/carbon dioxide co-electrolysis possesses stronger variations along with negative electrode thickness directions than those of the other two modes. This may be induced by the participation of more heterogeneous reactions with the reaction rate being two orders of magnitude higher in the H_2O/CO_2 co-electrolysis. It can be seen that the main zones of the heterogeneous reaction and the charge transfer reaction are located on the two sides of the negative electrode. When the negative electrode thickness is 700 μm, both the heterogeneous reaction and the charge transfer reaction do not occur within the

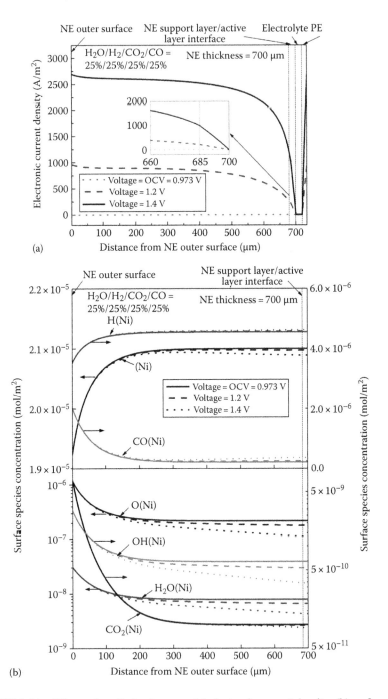

FIGURE 3.14 Effects of applied voltage on: (a) electronic current density; (b) surface species concentration. (Reprinted from *J. Power Sources*, 243, Li, W., Shi, Y., Luo, Y., Cai, N., Elementary reaction modeling of CO_2/H_2O co-electrolysis cell considering effects of cathode thickness, 118–130, Copyright 2013, with permission from Elsevier.)

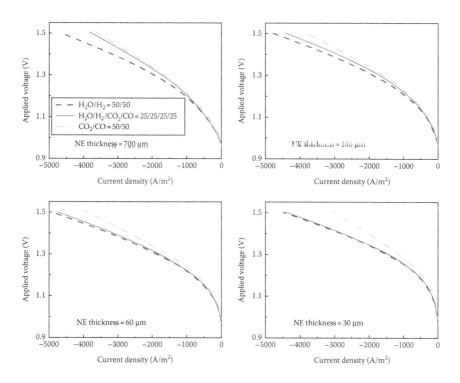

FIGURE 3.15 Polarization curves of three electrolysis modes with negative electrode thicknesses of 30–700 μm. (Reprinted from *J. Power Sources*, 243, Li, W., Shi, Y., Luo, Y., Cai, N., Elementary reaction modeling of CO_2/H_2O co-electrolysis cell considering effects of cathode thickness, 118–130, Copyright 2013, with permission from Elsevier.)

middle region of the negative electrode and, therefore, they cannot interact with each other. When the negative electrode thickness decreases however, both the chemistry active zone and the electrochemistry active zone overlap. Larger adsorbed O(Ni) coverage and fewer free sites (Ni) in the chemistry active zone can promote charge transfer reactions and further improve cell performances.

3.4.4.4 Active Zones for Heterogeneous Chemistry and Electrochemistry

The result divergence in negative electrode-supported and electrolyte-supported SOECs is well explained in Figure 3.18. At the negative electrode of H_2O/CO_2 co-electrolysis, there are two important reaction active zones: one is the heterogeneous chemistry active zone lying at the negative electrode outer surface, the other is the electrochemistry active zone locating near the negative electrode–electrolyte interface. The heterogeneous chemistry zone is two orders of magnitude more active than the electrochemistry zone. Hence, reactant gases tend to react heterogeneously when the two active zones overlap. When negative electrodes are hundreds of micrometers thick, the chemistry active zone and electrochemistry active zone would not overlap and heterogeneous chemical reactions and electrochemical reactions are isolated from with each other. Therefore, CO can be produced from both RWGS and

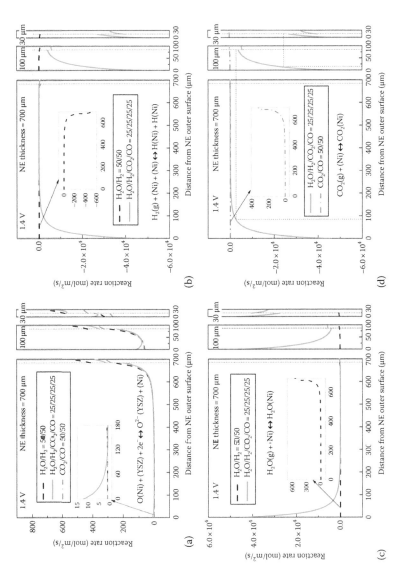

FIGURE 3.16 Reaction rates for (a) Charge transfer reaction and (b–h) heterogeneous elementary reaction of three electrolysis modes at 1.4 V with a negative electrode thickness of 30–700 μm. *(Continued)*

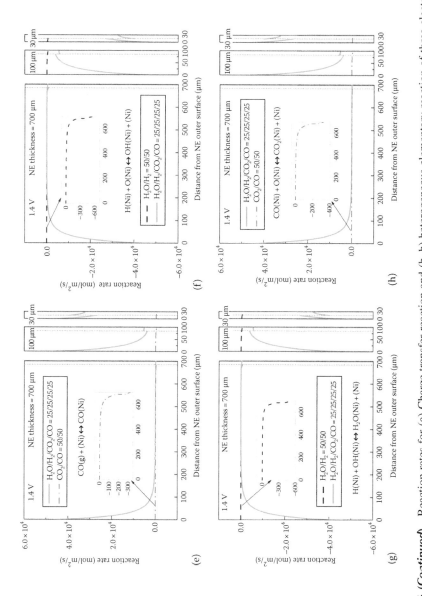

FIGURE 3.16 (Continued) Reaction rates for (a) Charge transfer reaction and (b–h) heterogeneous elementary reaction of three electrolysis modes at 1.4 V with a negative electrode thickness of 30–700 μm. (Reprinted from *J. Power Sources*, 243, Li, W. Shi, Y. Luo, Y. Cai. N.. Elementary reaction modeling of CO_2/H_2O co-electrolysis cell considering effects of cathode thickness, 118–130, Copyright 2013, with permission from Elsevier.)

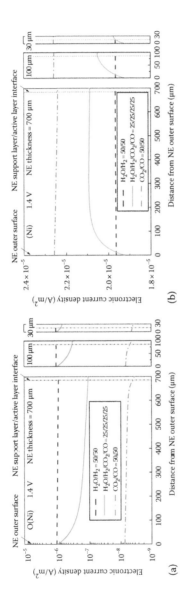

FIGURE 3.17 Concentration distributions of the surface species (a) O(Ni) and (b) (Ni) in three electrolysis modes at 1.4 V with negative electrode thicknesses of 30–700 μm. (Reprinted from *J. Power Sources*, 243, Li, W., Shi, Y., Luo, Y., Cai, N., Elementary reaction modeling of CO₂/H₂O co-electrolysis cell considering effects of cathode thickness, 118–130, Copyright 2013, with permission from Elsevier.)

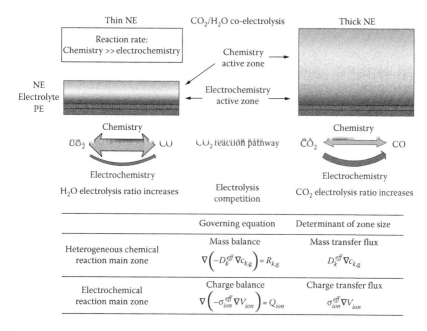

FIGURE 3.18 Heterogeneous chemistry/electrochemistry active zones, CO_2 reaction pathways, and mathematical expressions of the two zones. (Reprinted from *J. Power Sources*, 273, Li, W., Shi, Y., Luo, Y., and Cai, N., Elementary reaction modeling of solid oxide electrolysis cells: Main zones for heterogeneous chemical/electrochemical reactions, 1–13, Copyright 2015, with permission from Elsevier.)

CO_2 electrolysis. When the thicknesses of the negative electrode decreases to tens of micrometers, the two active zones would overlap. As heterogeneous chemical reactions are superior to electrochemical reactions, CO_2 mainly participates in heterogeneous chemical reactions in this case. Therefore, the electrochemical performances here are close to the performance under H_2O/H_2 atmospheres.

As mentioned before, the active zones for heterogeneous chemistry and electrochemistry can be affected by electrode structures and material parameters, such as conductivity, porosity, pore diameter, and tortuosity [58], as well as operating conditions, such as temperature and gas composition. The effects of these parameters on the size of the two active zones can be evaluated. First, the boundary of the two active zones needs to be clearly defined. The boundaries of the heterogeneous active zone are defined as the location where the concentration of any gaseous species is changed by 90% at OCV, while the boundaries of the electrochemistry active zone are defined as the location where the applied voltage is changed by 90% with an operation voltage of 1.4 V [61].

3.4.4.5 Effects of Porosity and Particle Diameter

Figure 3.19a through f shows the polarization curves with different porosities and particle diameters at 700°C with an oxidant/reductant ratio of 1.0. When the porosity ε_{NE} or particle diameter d_{NE} increases, the cell performance of the three electrolysis modes decreases. The ratio of steam to carbon dioxide participating in the electrolysis increases however with increasing ε_{NE} or d_{NE} values. Based on the particle coordination number theory in binary random packing of spheres as well as the percolation theory, the Ni surface area S_{Ni} and TPB area S_{TPB} are both in proportional to $(1-\varepsilon_{NE})/d_{NE}$ [61,62]. Therefore, the increase in porosity ε_{NE} or particle diameter d_{NE} reduces reaction active sites and decreases cell performance. Too much reduction in porosity and particle diameter is detrimental however due to gas and ion transport limitations [61,63].

Figure 3.19g and h shows the variation of chemistry and electrochemistry active zones with porosity and particle diameters [61]. According to the literature [61], increasing ε_{NE} can promote molecule and Knudsen diffusions. Therefore, the effective diffusion coefficient D_{eff} increases with increasing ε_{NE} and d_{NE} values, with ε_{NE} possessing a larger impact than d_{NE}. Increasing ε_{NE} can, however, on the other hand reduce effective conductivity; therefore, the electrochemistry active zone slightly decreases with increasing ε_{NE} values and slightly increases with increasing d_{NE} values. In summary, the increase of ε_{NE} or d_{NE} mainly affects the mass transfer process to enlarge the chemistry active zone, promoting CO production based on chemistry reactions and limiting CO_2 electrolysis. Therefore, the performance of H_2O/CO_2 co-electrolysis leaves from CO_2 electrolysis and approaches H_2O electrolysis.

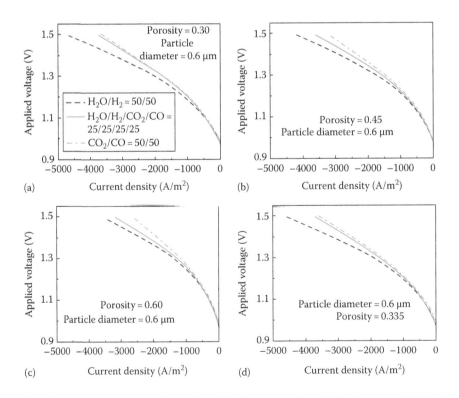

FIGURE 3.19 Effects of porosity and particle diameter on the polarization curves (a–f) and active zones of chemistry and electrochemistry (g, h).   (*Continued*)

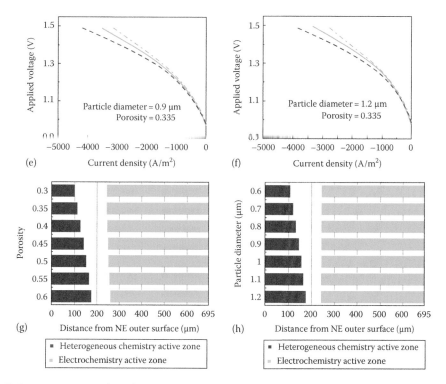

FIGURE 3.19 (*Continued*) Effects of porosity and particle diameter on the polarization curves (a–f) and active zones of chemistry and electrochemistry (g, h). (Reprinted from *J. Power Sources*, 273, Li, W., Shi, Y., Luo, Y., and Cai, N., Elementary reaction modeling of solid oxide electrolysis cells: Main zones for heterogeneous chemical/electrochemical reactions, 1–13, Copyright 2015, with permission from Elsevier.)

3.4.4.6 Effects of Ionic Conductivity

Figure 3.20a through c shows polarization curves with different ionic conductivities. When the ionic conductivity $\sigma_{ion,NE}$ increases, the electrochemical performances of the three electrolysis modes increase with growing differences, especially the performance of H_2O/CO_2 co-electrolysis that resembles the performance of CO_2 electrolysis. This is because $\sigma_{ion,NE}$ is independent of mass transfer but can greatly change the ohmic resistance. With increasing $\sigma_{ion,NE}$, electrochemistry reactions become more competitive and more CO_2 is reduced electrochemically. Figure 3.20d shows the variations of chemistry and electrochemistry active zones with ionic conductivity [61]. $\sigma_{ion,NE}$ affects electrochemistry active zones more but is independent of chemistry active zones.

3.4.4.7 Effects of Gas Composition and Temperature

Feed gas compositions have a significant effect on SOEC performances. As predicted [61], only 0.35% of CO_2 is electrolyzed when the oxidant to reactant ratio is 4:1. When the ratio of oxidant to reactant reaches 1:4 however, the fraction of CO_2

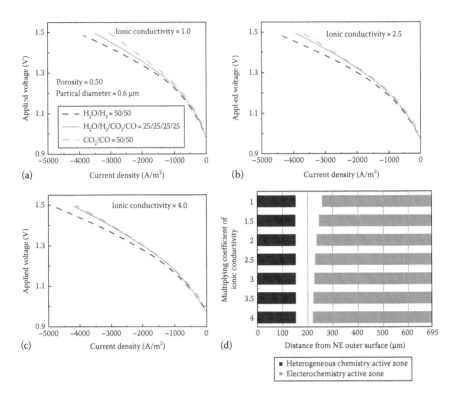

FIGURE 3.20 Effects of ionic conductivity on the polarization curves (a–c) and active zones of chemistry and electrochemistry (d). (Reprinted from *J. Power Sources*, 273, Li, W., Shi, Y., Luo, Y., and Cai, N., Elementary reaction modeling of solid oxide electrolysis cells: Main zones for heterogeneous chemical/electrochemical reactions, 1–13, Copyright 2015, with permission from Elsevier.)

electrolyzed increases to 12.03%. Figure 3.21a demonstrates that the size of the chemistry active zone does not increase linearly with increasing ratios of oxidant to reactant [61]. This is because the mass transfer flux $D^{eff}\nabla c$ is a governing variable of chemistry active zones. The value of ∇c also has great influences on chemistry active zones, in which $\sigma_{ion}^{eff}\nabla V_{ion}$ gradually increases with decreasing ratios of oxidant to reductant when the chemistry active zone location is close to the outer surface of the negative electrode. Thus, the size of the electrochemical active zone can significantly increase. Apart from transport process, gas compositions also influence both the heterogeneous chemistry active zone and the electrochemical active zone and electrochemical reactions are more significantly affected than heterogeneous chemical reactions. Therefore, the SOEC performance of H_2O/CO_2 co-electrolysis approaches CO_2 electrolysis when the ratio of oxidant to reductant is increased.

High temperatures are beneficial to both gas diffusion and ions conduction [61]. Therefore, the active zones of chemistry and electrochemistry will both increase with increasing temperatures as shown in Figure 3.21b. Temperatures do have a stronger

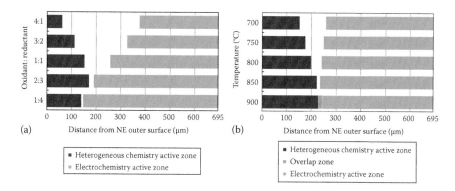

FIGURE 3.21 Effects of gas composition (a) and temperature (b) on the active zones of chemistry and electrochemistry. (Reprinted from *J. Power Sources*, 273, Li, W., Shi, Y., Luo, Y., and Cai, N., Elementary reaction modeling of solid oxide electrolysis cells: Main zones for heterogeneous chemical/electrochemical reactions, 1–13, Copyright 2015, with permission from Elsevier.)

effect on electrochemical reactions over other factors. The improvement to CO_2 electrolysis by rising temperature is more significant than that of H_2O electrolysis, therefore, the performance of H_2O/CO_2 co-electrolysis is closer to the performance of CO_2 electrolysis than to the performance of H_2O electrolysis.

3.4.5 Coupling of Reactions and Transfer Processes in Positive Electrodes

The structure of positive electrodes also has great effects on the cell performance of H_2O/CO_2 co-electrolysis. Here, the effects of porosity, pore diameters, and positive electrode thickness in both SOFC and SOEC modes are analyzed to provide an explanation for the transfer processes in positive electrodes.

Figure 3.22 shows the effects of positive electrode porosity ε_{PE} and pore diameter d_{PE} on overpotentials when being operated at the same current density absolute value in SOFC and SOEC modes with different positive electrode thicknesses [60]. A thickness of 100 μm is considered to be the optimal positive electrode thickness based on the overpotentials. The effects of the positive electrode thickness on the overpotential under SOFC mode are greater than that of SOEC mode. Under SOEC mode, the optimal porosity value is in the range of 0.25–0.3. When it comes to SOFC mode, the effects of ε_{PE} and positive electrode thickness are more significant than those on SOEC. The optimal porosity depends on the positive electrode thickness, with the optimal porosity in SOFC mode being 0.2 for a positive electrode thickness of 100 μm and the porosity has to be increased to 0.45 for a positive electrode thickness of 700 μm.

The effect of pore diameters on overpotentials becomes more significant when thickness of the positive electrode is less the 100 μm, due to inadequate reaction active sites in the thin positive electrode. In case of thick positive electrodes the ones with smaller pore diameters will have longer diffusion distances. Similar to

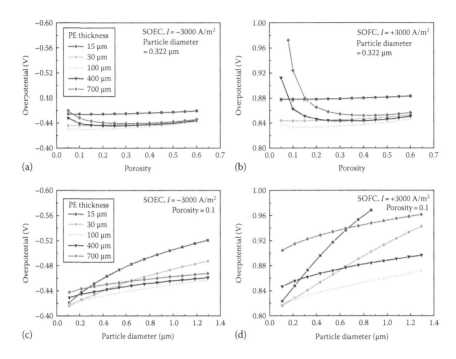

FIGURE 3.22 Effects of positive electrode porosity ε_{PE} (a,b) and pore diameter d_{PE} (c,d) on the overpotential for different positive electrode thicknesses in SOEC/SOFC modes. (Reprinted from *Int. J. Hydrogen Energy*, 39, Li, W., Shi, Y., Luo, Y., and Cai, N., Theoretical modeling of air electrode operating in SOFC mode and SOEC mode: The effects of microstructure and thickness, 13738–13750, Copyright 2014, with permission from Elsevier.)

the effects of porosity, pore diameters will impact the overpotential more in SOFC mode than in SOEC mode when the positive electrode thickness is greater than 100 μm.

Figure 3.23 shows the effects of positive electrode thickness (a), pore diameter d_{PE} (b), and positive electrode porosity ε_{PE} (c) on total positive electrode overpotentials and concentration overpotentials [60]. The increase in positive electrode thickness significantly increases concentration polarization, especially for SOFC mode. When the positive electrode thickness is 700 μm, the concentration polarization in the positive electrode of SOFC is 4.8-fold larger than the case of SOEC. Pore diameters have less impact on positive electrode concentration polarizations but more on ohmic and activation polarizations. With increasing positive electrode pore sizes, positive electrode concentration polarizations decrease in SOECs and increase in SOFCs. Porosity has significant influences on both the concentration polarization and total polarization in positive electrodes, especially for SOFC mode. The positive electrode concentration polarization decreases with increasing porosity. When ε_{PE} is less than or equal to 0.2 in SOFC mode, the concentration polarization exponentially increases with decreasing porosity.

3.4.6 Fuel-Assisted Electrolysis

Although SOECs have been demonstrated to be highly efficient, where a portion of its electricity demand (ΔG) can be compensated by its heat demand ($T\Delta S$) from elevated temperatures, it still requires large amounts of electricity to produce hydrogen/syngas at applicable operational temperatures. If this electricity is not obtained from renewable or abandoned sources but from other power sources, it will result in high operational costs. It can be three or four times more expensive than that of steam–methane reforming [66]. Therefore, it is necessary to decrease the electricity

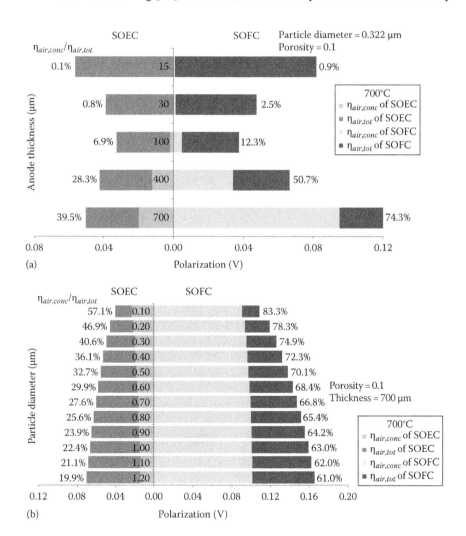

FIGURE 3.23 Effects of positive electrode thickness (a), pore diameter d_{PE} (b), and positive electrode porosity ε_{PE}. *(Continued)*

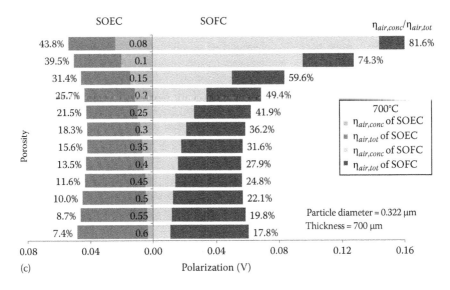

(c)

FIGURE 3.23 (*Continued*) Effects of positive electrode thickness (c) on the total positive electrode overpotential and concentration overpotential. (Reprinted from *Int. J. Hydrogen Energy*, 39, Li, W., Shi, Y., Luo, Y., and Cai, N., Theoretical modeling of air electrode operating in SOFC mode and SOEC mode: The effects of microstructure and thickness, 13738–13750, Copyright 2014, with permission from Elsevier.)

consumption of SOECs. In recent years, there is growing interest in solid oxide fuel–assisted electrolysis cells (SOFECs). Unlike SOECs, fuel (such as carbon, CO, CH$_4$, biomass, or other hydrocarbon fuels) is sent into the positive electrode to facilitate the removal of oxygen in SOFEC mode. Figure 3.24 demonstrates the operating principles of H$_2$O electrolysis by SOFEC and SOEC. The half-cell reaction in the negative electrodes of both SOECs and SOFECs is

$$H_2O + 2e^- \rightarrow H_2 + O^{2-} \tag{3.32}$$

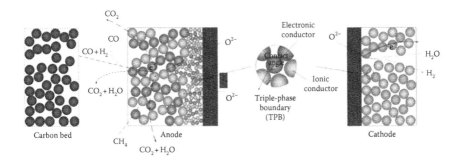

FIGURE 3.24 Basic working principles of SOFECs for H$_2$O electrolysis [64,65]. (From Luo, Y. et al., *Int. J. Hydrogen Energy*, 39, 10359, 2014; Wu, Y. et al., *Solid State Ionics*, 295, 78, 2016.)

The half-cell reaction in the positive electrode of SOECs is

$$O^{2-} \rightarrow \frac{1}{2}O^2 + 2e^- \tag{3.33}$$

As for SOFECs using CO, CH_4, and carbon as assisting-fuels, the half-cell reactions in the positive electrode (when fully oxidized) are as follows:

$$CO + O^{2-} \rightarrow CO_2 + 2e^- \tag{3.34}$$

$$\frac{1}{4}CH_4 + O^{2-} \rightarrow \frac{1}{2}H_2O + \frac{1}{4}CO_2 + 2e^- \tag{3.35}$$

$$C + CO_2 \rightarrow 2CO$$
$$CO + O^{2-} \rightarrow CO_2 + 2e^- \tag{3.36}$$

The total cell reactions of SOECs and SOFECs are expressed as follows:

$$\text{SOEC: } H_2O \rightarrow \frac{1}{2}O^2 + H_2 \tag{3.37}$$

$$\text{CO-SOFEC: } CO + H_2O \rightarrow CO_2 + H_2 \tag{3.38}$$

$$CH_4\text{-SOFEC: } \frac{1}{2}H_2O + \frac{1}{4}CH_4 \rightarrow \frac{1}{4}CO_2 + H_2 \tag{3.39}$$

$$\text{Carbon-SOFEC: } \frac{1}{2}C + H_2O \rightarrow \frac{1}{2}CO_2 + H_2 \tag{3.40}$$

It can be figured out that the working principles of SOFECs are actually an electrochemical reforming reaction. Thermodynamically, adding assisting-fuels can significantly reduce the reversible potential (Nernst potential E_N) of chemical energy by fuel oxidation [67,68]. The theoretical electricity demands of steam electrolysis through SOECs and SOFECs in a temperature range of 100°C–1000°C are shown in Figure 3.25. OCV is determined by the Nernst potential difference between the two electrodes, representing the minimum electrical demand (at zero current). The theoretical reversible voltage of SOFECs is at least 1.0 V lower than that of SOECs. Typical polarization curves of SOECs and SOFECs are drawn in Figure 3.26 [64]. The power density (P, W/m²) of the cell is expressed as

$$P = iV_{cell} \tag{3.41}$$

where
i denotes the current density (A/cm²)
V_{cell} denotes the cell voltage (V)

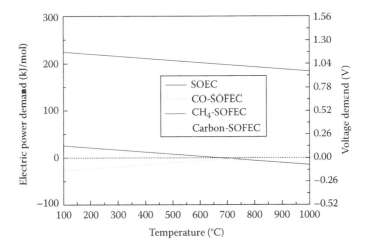

FIGURE 3.25 Electric power demands of H_2O electrolysis by SOEC and SOFEC. (From Luo, Y. et al., *Int. J. Hydrogen Energy*, 39, 10359, 2014.)

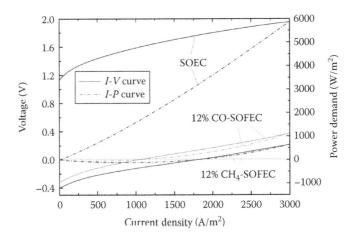

FIGURE 3.26 I-V curves and I-P curves of SOECs and SOFECs. (Reprinted from *Int. J. Hydrogen Energy*, 39, Luo, Y., Shi, Y., Li, W., Ni, M., and Cai, N., Elementary reaction modeling and experimental characterization of solid oxide fuel-assisted steam electrolysis cells, 10359–10373, Copyright 2014, with permission from Elsevier.)

If the current density in the electrolysis of pure water (SOEC) is defined as a positive value (current flows from the positive electrode, i.e., anode electrode where O_2 formation occurs, to the negative electrode, i.e., cathode electrode where H_2 formation occurs), the cell voltage should also be a positive value. If the current density is negative, the cell will operate as an SOFC. As observed, the V_{cell} of SOFECs is generally negative. In this case, the power density (Equation 3.41) is negative, meaning that the SOFEC can produce gaseous products and electricity simultaneously. However, if the current density is high enough, V_{cell} will become positive, meaning that much less additional electricity is required to accelerate H_2 generation rates. In general, for the same H_2 generation rate, the ratio of electricity saved by SOFEC virus the electricity consumed by SOECs can be calculated as follows [3]:

$$\frac{E_{SOEC} - E_{SOFEC}}{E_{SOEC}} = \frac{V_{SOEC} - V_{SOFEC}}{V_{SOEC}} \qquad (3.42)$$

Therefore, moderate current densities can be achieved with minimal electrical energy consumption by adding cheaper assisting-fuels.

Although electrical energy demands of SOFECs are significantly smaller, efficiency analysis is still required. The involved energy flows are summarized in Figure 3.27a, and the results are shown in Figure 3.27b. Here, the figure demonstrates that when irreversible heat is offer to fulfill the reversible heat demands (except for the CO-SOFEC case), efficiency is indeed improved by introduction of assisting-fuels.

Among all the potential assisting-fuels, carbon or carbonaceous solids offer several advantages over other gaseous fuels. These advantages include higher abundance of raw materials, cost effectiveness, and lower quality requirements. These carbon or carbonaceous solid fuels including coal and biomass possess higher theoretical efficiencies and higher CO_2 emission reduction potentials. C-assisted SOFECs have the potential to produce syngas at both electrodes (producing CO at the anode and hydrogen at the cathode) through the easy control of H_2/CO ratios. This is helpful in subsequent processes to synthesize other chemicals from the syngases produced [65,69,70]. The net production rates of CO and H_2 as a function of current density in a button carbon-SOFEC are shown in Figure 3.28.

3.5 OPERATING CONDITION DESIGNS AND DYNAMIC BEHAVIORS IN TUBULAR CELLS

Tubular cells are common configurations of commercialized SOECs. They possess advantages such as feasible sealing, easy scalability, good mechanical strength, and high thermal shock resistance compared with planar cells [18,71]. However, the distributions of current densities, gas species concentrations, and temperatures inevitably become more nonuniform with increasing reaction areas. Therefore, it is necessary to describe the distributions within tubular cells and optimize operating conditions to provide better cell performance for both steady and dynamic operations.

3.5.1 EXPERIMENT

In experiments, the tested tubular SOEC included dead-end and inner-negative electrodes with dimensions of 5 mm inner diameter, 6.59 mm outer diameter, 115 mm of length, and 70 mm of effective length, as shown in Figure 3.29 [72]. Figure 3.29 also demonstrates the polarization curves of the tested tubular SOEC in a temperature range of 550°C–650°C with a negative electrode inlet flow rate of 100 mL/min and a positive electrode inlet flow rate of 200 mL/min. If no hydrogen is introduced, the current at the

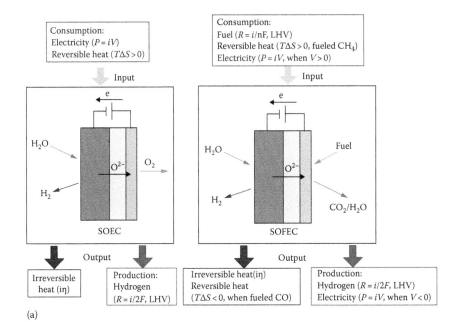

(a)

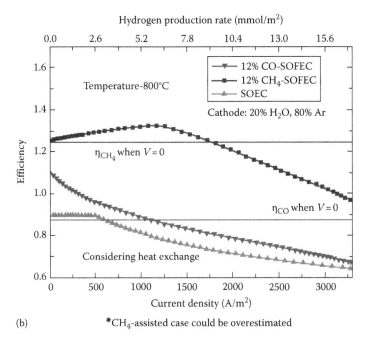

(b)   *CH_4-assisted case could be overestimated

FIGURE 3.27 Efficiency analysis of the comparison between (a) SOECs and (b) SOFECs. (Reprinted from *Int. J. Hydrogen Energy*, 39, Luo, Y., Shi, Y., Li, W., Ni, M., and Cai, N., Elementary reaction modeling and experimental characterization of solid oxide fuel-assisted steam electrolysis cells, 10359–10373, Copyright 2014, with permission from Elsevier.)

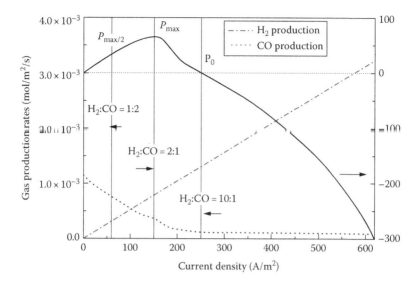

FIGURE 3.28 The net production rates of CO and H_2 as a function of current density by carbon-SOFECs. (Reprinted from *Solid State Ionics*, 295, Wu, Y., Shi, Y., Luo, Y., and Cai, N., Elementary reaction modeling and experimental characterization of solid oxide direct carbon-assisted steam electrolysis cells, 78–89, Copyright 2016, with permission from Elsevier.)

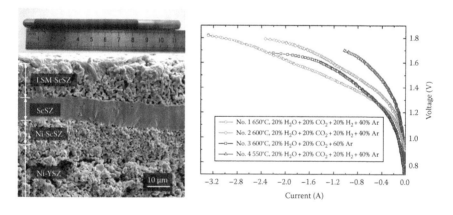

FIGURE 3.29 Photo, SEM image, and polarization curves of a tubular SOEC. (Reproduced from Luo, Y. et al., *J. Electrochem. Soc.*, 162, F1129, Copyright 2015, The Electrochemical Society.)

same applied voltage is higher. The outlet gas composition for each condition is collected and measured using a gas chromatograph at both OCV and 1.5 V [72]. When the negative electrode is fed with a combination of 20% H_2O, 20% CO_2, 20% H_2, and 40% Ar, RWGS and methanation occur to produce CO and CH_4 even without electrical input. When electricity is applied, electrochemical reaction rates are promoted for CO_2 conversion. CH_4 production is reduced when hydrogen feeding is decreased, corresponding to a methanation equilibrium. Thermodynamics predict that lower temperatures will promote the methanation reactions. Thus, CH_4 production rates can be enlarged to 9.46% at 550°C with the addition of H_2 even when no electricity is applied and can reach up to 12.34% at 1.5 V. CH_4 production rates at temperatures of 550°C, 600°C, and 650°C are all increased by 3%–4% when a voltage of 1.5 V is applied.

3.5.2 COMPREHENSIVE ELECTROTHERMAL MODEL FOR TUBULAR SOECs

A two-dimensional comprehensive electrothermal dynamic model has been built to couple heat transfer, charge transfer, fluid flow, gas diffusion, and chemical/electrochemical reactions within a tubular SOEC, as shown in Figure 3.30 [73,74]. This model has been validated through measured data from tested tubular SOECs.

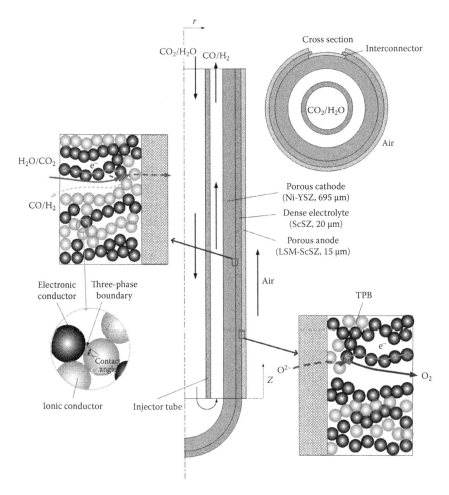

FIGURE 3.30 Photo, SEM image, and polarization curves of a tubular SOEC. (Reprinted from *Energy*, 70, Luo, Y., Shi, Y., Li, W., and Cai, N., Comprehensive modeling of tubular solid oxide electrolysis cell for co-electrolysis of steam and carbon dioxide, 420–434, Copyright 2014, with permission from Elsevier.)

The distributions of current density, temperature, gas concentration, and velocity are demonstrated in Figure 3.31 [73,74]. As the figure presents, concentrations of CO and H_2 both increase along with gas flow. The temperature fields show that thermal effects depend on applied voltages. In an atmosphere composed of 40% H_2O, 40% CO_2, 10% H_2, and 10% CO, the tubular SOEC is thermal-neutral at 1.33 V, endothermal below 1.33 V, and exothermal above 1.33 V. In Figure 3.31a, the highest current density found to be at the middle of the tube, corresponding to the "hottest" point of the SOEC in the exothermic mode.

3.5.2.1 Optimization of Conversion Ratio and Efficiency

Fuel production and efficiency are two key parameters that presents performance in SOECs. The conversion ratios ξ_{H_2}, ξ_{CO}, ξ_{syn} of H_2, CO, and syngas produced from H_2O/CO_2, as well as the efficiency η, can be defined respectively as

$$\xi_{H_2} = \frac{q_{H_2}^{in} - q_{H_2}^{out}}{q_{H_2O}^{in}}, \quad \xi_{CO} = \frac{q_{CO}^{in} - q_{CO}^{out}}{q_{CO_2}^{in}}, \quad \xi_{syn} = \frac{q_{H_2}^{in} - q_{H_2}^{out} + q_{CO}^{in} - q_{CO}^{out}}{q_{H_2O}^{in} + q_{CO_2}^{in}} \quad (3.43)$$

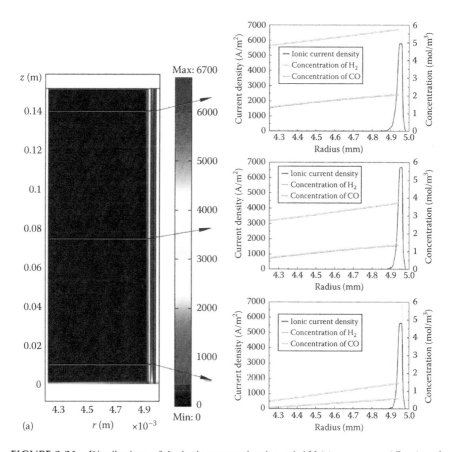

FIGURE 3.31 Distributions of the ionic current density at 1.4 V (a). *(Continued)*

$$\eta = \frac{q_{H_2}^{out} LHV_{H_2} + q_{CO}^{out} LHV_{CO}}{IV + q_{NE}^{in} M_{NE} \int_{298}^{T_{in}} C_{p,ca} dT + q_{PE}^{in} M_{PE} \int_{298}^{T_{in}} C_{p,an} dT} \tag{3.44}$$

where

q represents molar flow rates (mol/s)

LHV is the lower heating value

ξ_{H_2}, ξ_{CO}, ξ_{syn}, and η, expressed by Equations 3.43 and 3.44 at different temperatures, gas compositions, applied voltages, and gas flow velocities, are shown in Figure 3.32. Figure 3.32b shows that higher concentrations of H_2O can significantly

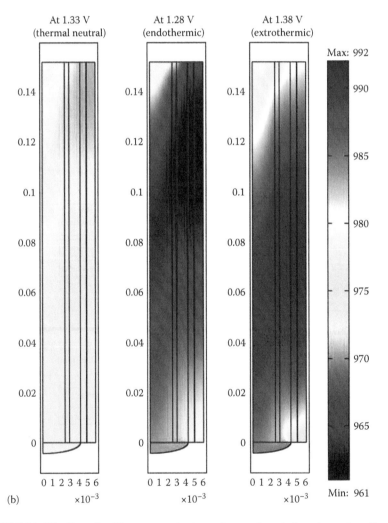

(b)

FIGURE 3.31 (Continued) Temperature in thermal-neutral, endothermic, and exothermic modes (b). *(Continued)*

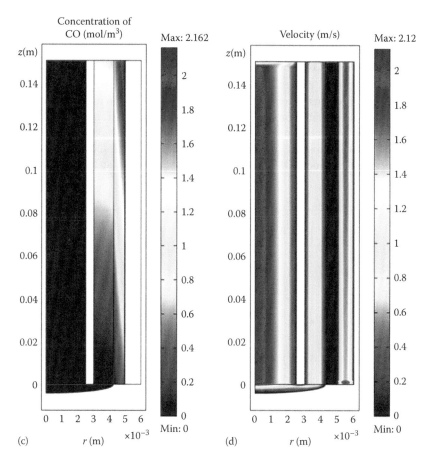

FIGURE 3.31 (*Continued*) CO concentration (c); and gas flow velocity (d). (Reprinted from *Energy*, 70, Luo, Y., Shi, Y., Li, W., and Cai, N., Comprehensive modeling of tubular solid oxide electrolysis cell for co-electrolysis of steam and carbon dioxide, 420–434, Copyright 2014, with permission from Elsevier; Reprinted from *Energy*, 89, Luo, Y., Shi, Y., Li, W., and Cai, N., Dynamic electro-thermal modeling of co-electrolysis of steam and carbon dioxide in a tubular solid oxide electrolysis cell, 637–647, Copyright 2015, with permission from Elsevier.)

improve syngas conversion ratios and cell performances but will slightly reduce ξ_{H_2} and ξ_{CO}. Lower velocities, meaning lower inlet gas flow rates, can increase both conversion ratios and efficiencies but decrease fuel yields. Moreover, if gas flow rates are too low, downstream concentration overpotentials could dramatically increase due to the consumption of reactants. Therefore, if the inlet flow velocity is chosen to be 1.0 m s^{-1}, the optimal operating conditions for tubular SOECs are H$_2$O/CO$_2$ molar ratio of 1.0, 700°C, and 1.4 V. In this case, efficiencies of 59.4% and syngas conversion ratios of 43.8% can be achieved. Here, the relatively low efficiency level is due to the fact that the energy used to heat the inlet gas is included in the total energy consumption of the system when calculating the efficiency. In practical operation, exhaust heat from SOEC outlet gases or other industrial waste heat sources can be utilized for this heat requirement through means of heat exchangers.

3.5.2.2 Operating Condition Designs for Methane Production

The models discussed earlier can be employed to evaluate operating conditions such as the distributions of temperature and chemical reaction rates. Figure 3.33a and b shows the distributions of WGSR and methanation rates within porous negative electrodes operating in co-flow mode [72]. Figure 3.33c displays the temperature field in co-flow and counterflow modes [72]. In co-flow mode, the temperature in the middle of the tube is the highest, being at least 50°C higher

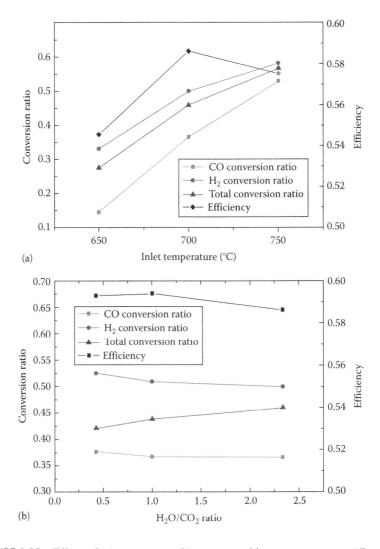

(a)

(b)

FIGURE 3.32   Effects of (a) temperature, (b) gas composition.   (*Continued*)

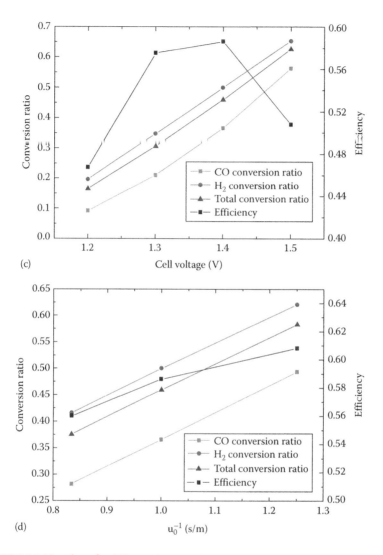

FIGURE 3.32 (Continued) Effects of (c) applied voltage, and (d) inlet velocity on gas conversion ratio and efficiency. (Reprinted from *Energy*, 70, Luo, Y., Shi, Y., Li, W., and Cai, N., Comprehensive modeling of tubular solid oxide electrolysis cell for co-electrolysis of steam and carbon dioxide, 420–434, Copyright 2014, with permission from Elsevier.)

than at the downstream location. The inlet gas composition leads to WGSR proceeds in the reverse direction (RWGS) to produce CO. When CO accumulates however, the WGSR goes in the forward direction. Active zones of methanation rates are located in the low-temperature regions; therefore, the downstream location has more significant methanation, whereas the upstream location is a more crucial zone for CO_2/H_2O co-electrolysis to provide enough methanation reactants for the downstream. To achieve this scenario, thermal distributions with high-temperature upstream and low-temperature at downstream are required. The counterflow mode can form such a thermal distribution where temperature differences between the upstream and downstream locations can reach roughly 100°C. For example, corresponding distributions of CH_4 production ratios in these two flow modes are shown in Figure 3.33d [72]. Local CH_4 production ratios at the downstream location in counterflow mode surpass that in co-flow mode. The counterflow mode can also improve the total current by 14.3%. Consequently, outlet CH_4 production ratios are increased by at least 2% from flow designs. Enlarging applied voltage and temperature differences can increase the effects of this flow design but may be harmful for SOEC durability.

Figure 3.34 shows the effects of steam content and pressure on CH_4 production. At atmospheric pressure, CH_4 production ratios increase significantly by decreasing the steam content. CH_4 production ratios with 10% H_2O can reach 12.5% at 1.5 V in the counterflow mode. CH_4 production ratios are far more sensitive to

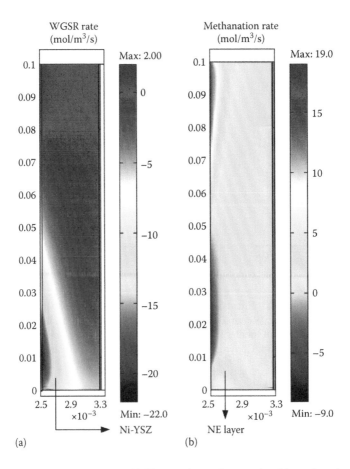

FIGURE 3.33 Distributions: (a) WGSR rates in co-flow mode; (b) methanation rates in counterflow mode. *(Continued)*

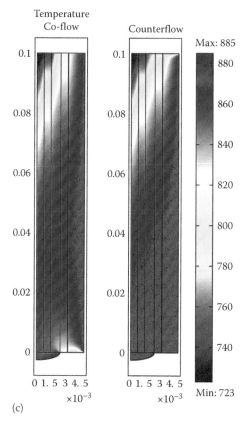

(c)

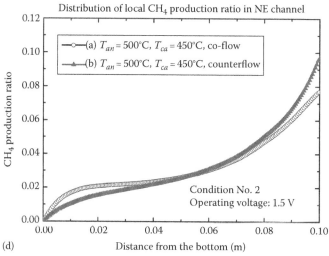

(d)

FIGURE 3.33 (Continued) Distributions: (c) temperature in co-flow and counterflow modes; (d) local CH$_4$ production ratios of two flow modes in negative electrode channel. (Reproduced from Luo, Y. et al., *J. Electrochem. Soc.*, 162, F1129, Copyright 2015, The Electrochemical Society.)

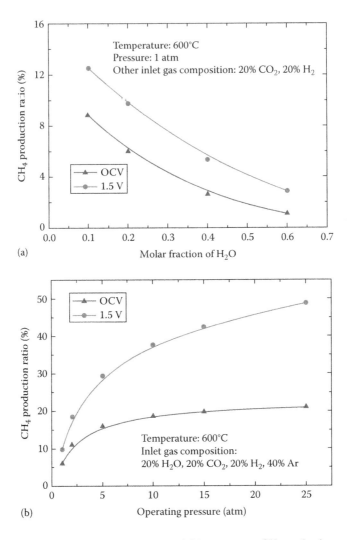

FIGURE 3.34   Effects of (a) steam content and (b) pressure on CH_4 production.

pressurizations. At 25 atm, CH_4 production ratios can reach 49.8% at 1.5 V. This methane-producing, one-step conversion process through SOECs is high enough to support power-to-gas technologies to play a part in renewable energy storage applications.

3.5.2.3   Dynamic Operation Behaviors

When SOECs are used for renewable energy storage purposes, intermittence and fluctuations in renewable energy sources cause dynamic conditions in SOEC operations, especially in the case of transient electricity input. To study these dynamic effects, temperature, gas concentration, and total current of SOECs are selected to represent the dynamic processes of heat transfer, mass transfer, and charge transfer respectively, as displayed in Figure 3.35. The simulation results show that tubular SOEC performances take roughly 2000 s to reach a new steady state. The difference between the thermal case and isothermal case in Figure 3.35a indicates that this slow change in thousands of seconds is attributed to slow heat transfer. The isothermal case shows that tubular SOECs take about 1 s to reach a new steady state, corresponding to fast variations in the thermal case. An overshoot in the curve of the transient current reveals that charge transfer is faster than mass transfer. To identify the response times of these three transfers, a time constant τ is defined as the time a key parameter spends to reach 90% of the change value between the two steady states. The τ values of charge transfer and mass

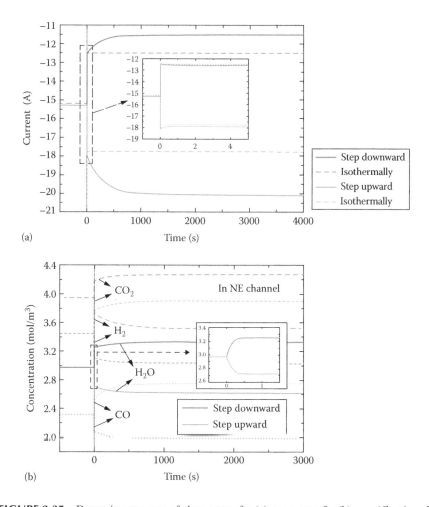

(a)

(b)

FIGURE 3.35   Dynamic responses of charge transfer (a), mass transfer (b).   (*Continued*)

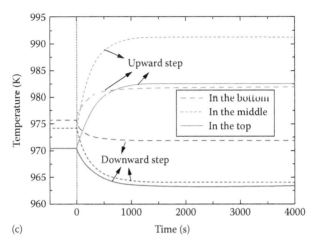

(c)   Time (s)

FIGURE 3.35 (*Continued*)   Heat transfer (c). (Reprinted from *Energy*, 89, Luo, Y., Shi, Y., Li, W., and Cai, N., Dynamic electro-thermal modeling of co-electrolysis of steam and carbon dioxide in a tubular solid oxide electrolysis cell, 637–647, Copyright 2015, with permission from Elsevier.)

transfer are obtained from the isothermal case. The values of the time constants are calculated to be $\tau_E = 0.011$ s for charge transfer, $\tau_M = 0.255$ s for mass transfer, and $\tau_T = 515$ s for heat transfer.

Figure 3.36 shows the responses of transient efficiencies η, ξ_{H_2O}, ξ_{CO_2}, ξ_{syn}, and molar ratios of generated H_2/CO $\varphi_{H_2/CO}$ to voltage steps, inlet gas concentrations, and inlet temperatures [74]. The efficiencies observed appear to suddenly increase and then slowly decrease when voltage steps are downward, inlet H_2O content steps are upward (inlet CO_2 content steps down), or inlet temperature steps are downward. When the steps are reversed, the efficiency decreases suddenly and then increases slowly. Similarly, ξ_{syn} increases suddenly and then decreases slowly with an upward inlet H_2O content (a downward inlet CO_2 content) or a downward inlet temperature. Therefore, dynamic operations can be designed to fully utilize the different time delays of the charge/mass/heat transfer process to improve efficiencies by introducing sudden increases but then avoiding slow decreases of efficiency. A cyclic voltage which decreases in milliseconds and increases in seconds can improve the time-average efficiency. As for the dynamic operation of inlet gases, charge transfers are fast enough to be ignored. The inlet H_2O content, increasing in kiloseconds and then decreasing in seconds (inlet CO_2 content decreasing in kiloseconds and increasing in seconds), is reasonable enough to promote time-average efficiencies and conversion ratios, whereas for the dynamic operation of inlet gas temperatures, the heat transfer is slow enough to neglect the dynamic behaviors of charge transfer and mass transfer. The dynamic operation of inlet temperatures can cause significant impacts on tubular SOECs due to thermal shocks, but it can also significantly stabilize temperature fluctuations caused by other dynamic behaviors.

3.6 HIGH-TEMPERATURE ELECTROLYSIS SYSTEMS AND INTEGRATION WITH RENEWABLE/FOSSIL ENERGY SYSTEMS

3.6.1 System Integration and Typical Configurations

In practical applications, SOECs integrate with various key components to utilize electrical power and produce gaseous/liquid fuels [3,75,76]. An energy storage system using SOEC technology usually contains a power generation device, user loads and functional components for electrolysis, thermal management, power conditioning, reactant supply, product post-processing, and so on [3,77]. SOEC system designs should account for the integration of various components, the matching of power and heat, and the reduction of system costs in applications. Cost reduction and

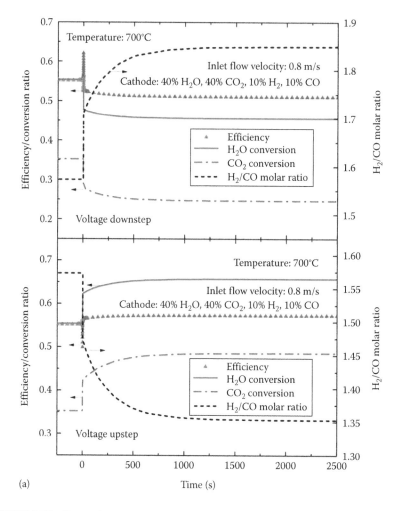

FIGURE 3.36 Dynamic responses of the efficiency and conversion ratio to voltage steps (a).

(Continued)

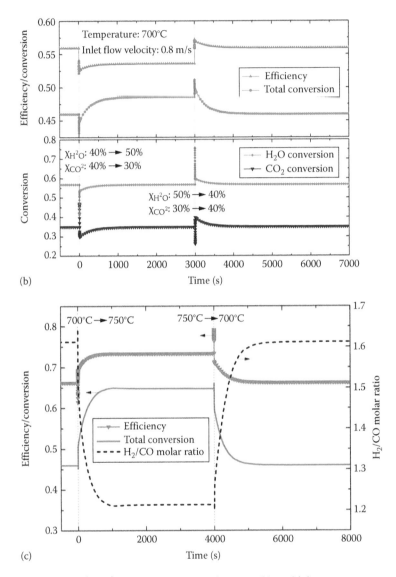

FIGURE 3.36 (*Continued*)   Inlet gas concentration steps (b), and inlet temperature steps (c). (Reprinted from *Energy*, 89, Luo, Y., Shi, Y., Li, W., and Cai, N., Dynamic electro-thermal modeling of co-electrolysis of steam and carbon dioxide in a tubular solid oxide electrolysis cell, 637–647, Copyright 2015, with permission from Elsevier.)

component integration optimization often conflict with each other and compromises are often needed. Therefore, system designs have to take economic and technological feasibilities into consideration.

Due to operating temperature ranges of 500°C–1000°C in SOEC systems, thermal management is necessary to ensure full utilization of system heat. This generally includes inlet gas preheating, heat recuperation, and outlet gas cooling.

Due to the complex processes related to heat release or absorption within SOEC stacks, temperatures within SOEC stacks have to be controlled to prevent thermal cracking, especially in the application of intermittent and fluctuating renewable power sources such as wind or solar power. As mentioned before, the heat produced from various polarizations can be used to support endothermic electrolysis reactions of H_2O and CO_2, even reaching a thermal neutral state. A usual method to regulate stack temperatures is to adjust the airflow rate and temperature. The Idaho National Lab built a 15 kW integrated planar SOEC facility and set up a system simulation platform to simulate syngas production driven by nuclear energy. They predicted an optimal power-to-syngas efficiency of 48.3% [3,78,79].

When SOECs are applied in reversible operations as a reversible solid oxide cell (RSOC), generated hydrogen, syngas, and even methane can be directly fed into RSOC stacks in SOFC operation mode. Aiming for high-yield fuel productions, a product post-processing system is required to generate targeted hydrocarbon fuels through further methanation or F-T synthesis [3]. This product post-processing is suitable for domestic and industrial applications.

A power conditioning module is crucial to stabilize the applied voltages of SOECs, especially for renewable energy applications. To increase durability, SOEC units require direct current inputs at relatively stable voltages, which is one of the purposes of adopting a power conditioning subsystem [3,80].

Figure 3.37 shows representative system configurations for H_2O electrolysis and H_2O/CO_2 co-electrolysis driven by renewable power or nuclear power. If the inlet H_2O is in a liquid phase, large amount of heat is required [3]. Thus, heat resources and steam exhaust by-products, such as exhausted gases in power plants or chemical plants, are beneficial to the economic performance of the system. Because of this, researchers are paying more attention to the integration of SOECs with nuclear reactors due to its high stability and abundant steam supply. Renewable power storage systems using SOECs are also gaining attention in the application of improving renewable penetration and system stability, especially for the mitigation of inconsistencies in wind or solar power sources.

In regard to H_2O/CO_2 co-electrolysis systems, configurations are more complex due to the introduction of CO_2 and the processing of syngases. These are advantages to the systems however as they reduce CO_2 emissions and produce syngas. CO_2 is generally captured from power plants and coal chemical plants through precombustion CO_2 capture, post-combustion CO_2 capture, oxy-fuel combustion process, and even air-capture or seawater-capture technologies [3,81]. The most mature technologies in utilizing syngas products are F-T synthesis and methanation, the F-T synthesis process has been widely used in the commercial production of gasoline and diesel. The simplified reaction of F-T synthesis is expressed as [3,82]

$$CO + 2H_2 \rightarrow [-CH_2-] + H_2O - 165\,kJ/mol \qquad (3.45)$$

The methanation process, as mentioned earlier, can convert H_2 and CO to methane when catalyzed by nickel, iron, or cobalt catalysts [3,83]. Due to existing natural gas networks, methanation reactors integrated with SOECs can be a feasible

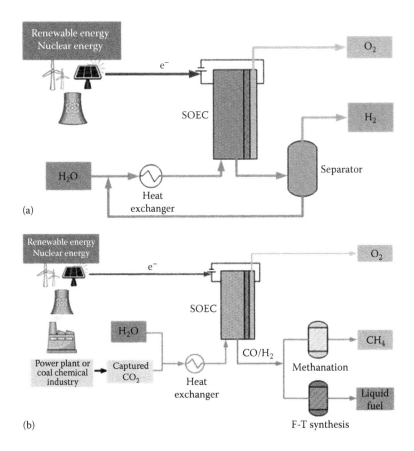

FIGURE 3.37 Featured system configurations for hydrogen (a) and syngas production (b) by SOECs. (Shi, Y., Luo, Y., Li, W., Ni, M., and Cai, N.: High temperature electrolysis for hydrogen or syngas production from nuclear or renewable energy. Yan, J. (ed.), *Handbook of Clean Energy Systems*. New York. 2015. Copyright Wiley-VCH Verlang GmbH & Co. KGaA. Reproduced with permission.)

energy storage technology. Using a promising potential application as an example, Figure 3.38 demonstrates an integration of H_2O/CO_2 co-electrolysis with coal-to-methane systems [3]. In existing commercialized coal-to-methane processes, H_2O and CO_2, the reactants of SOECs, can be produced during methanation and WGSR processes. Therefore, appropriate locations for such a system are in areas where coal chemical industries and renewable power sources are in close proximity to each other. A promising bidirectional connection between renewable energies and natural gases can be built through the application of SOECs [84]. A combined study carried out by the Technical University of Denmark (DTU), the Colorado School of Mines, and Northwestern University demonstrated a Power-to-Gas (PtG) system through RSOC with subsurface storage of CO_2 and CH_4. This system was predicted to reach round-trip efficiencies exceeding 70% and storage costs of approximately $3¢ \cdot kW^{-1} \cdot h^{-1}$ [85]. A renewable natural gas hybrid system combining wind power,

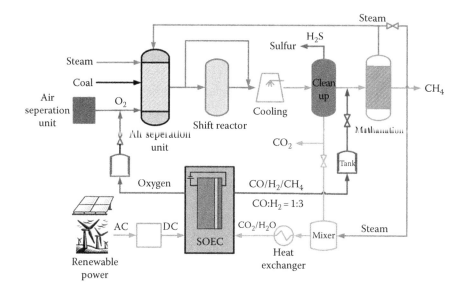

FIGURE 3.38 Schematic diagram of an SOEC utilized for methane production from renewable power and coal. (Shi, Y., Luo, Y., Li, W., Ni, M., and Cai, N.: High temperature electrolysis for hydrogen or syngas production from nuclear or renewable energy. Yan, J. (ed.), *Handbook of Clean Energy Systems*. New York. 2015. Copyright Wiley-VCH Verlang GmbH & Co. KGaA. Reproduced with permission.)

solar power, gas internal combustion engine (GICE), RSOC, and lithium-ion battery is shown on Figure 3.39. The simulation platform of this system was built using the gPROMS platform, and the effects of renewable power penetration were evaluated when RSOC alone is applied as the power storage device. As identified, the mismatch between the power supply and the demand increased with increasing renewable power penetration.

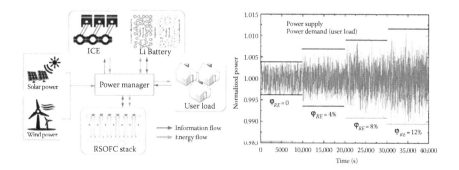

FIGURE 3.39 System configuration of a renewable energy and natural gas hybrid system and its simulation platform using the gPROMS platform [86]. (Reprinted from *J. Power Sources*, 340, Luo, Y., Shi, Y., Zheng, Y., and Cai, N., Reversible solid oxide fuel cell for natural gas/renewable hybrid power generation systems, 60–70, Copyright 2017, with permission from Elsevier.)

3.6.2   NOVEL CRITERIA FOR RENEWABLE POWER STORAGE SYSTEMS

In renewable power storage systems, matching power supply and demand is important in evaluating power quality and system dynamic stability of operation. The difference between power supply and demand varies with time however, with no satisfactory quantifying criteria. A novel criterion, mutual information, can be proposed to identify the effects of renewable power penetration and power storage capacity. Mutual information (MI) is a measurement of information involving information theory and probability theory, essentially the combination of information entropies of two variables. For example, MI can measure the general dependence of power supply and demand when compared with other criteria such as maximum relative deviation or correlation [87–91]. When MI is applied to quantify the matching of power supply and demand, the value of MI $I(S,D)$ and its derivation from the Shannon entropy can be expressed as Equations 3.46 and 3.47 [91–94]:

$$I(S, D) = \sum_D \sum_S P(S, D) \ln \frac{P(S, D)}{P(S)P(D)} \tag{3.46}$$

$$s(S) = -\sum_S P(S) \ln P(S)$$

$$s(D) = -\sum_D P(D) \ln P(D)$$

$$s(S, D) = -\sum_D \sum_S P(S, D) \ln P(S, D) \tag{3.47}$$

$$I(S, D) = s(S) + s(D) - s(S, D) \geq 0$$

where
  s is the Shannon entropy
  $s(S,D)$ is the joint entropy of power supply and demand

The two applications of MI in renewable power storage systems are the evaluation of the impact of renewable power penetration and the detection of power storage capacity. The mismatch between power supply and demand in different time points becomes larger with increasing renewable power penetration φ_{RE} when RSOC is applied as the power storage device. Correspondingly, the value of MI decreases proportionately with increasing φ_{RE}, and the decrease becomes slower with increasing φ_{RE}. Lithium-ion battery power storage displays a higher MI than RSOC power storage due to the better dynamic behaviors of lithium-ion batteries. Furthermore, the MI in the case of lithium-ion batteries maintains high sensitivity to variations of φ_{RE} even at large φ_{RE}. This phenomenon implies that besides φ_{RE} levels, MI can also reflect the dynamic responses of a power device.

The other important application of MI is to detect the suitable capacity of a power storage device. Using lithium-ion batteries as an example, when the capacity of a Li-ion battery declines to below the critical capacity, the MI value drops suddenly from 3.89 to 0.0016. This phenomenon reveals that MI is sensitive to critical capacity even if the power supply deviates from the power demand at only one time point. Similarly, this method can also be used to detect the capacity of other power devices, such as wind generators, GICEs, and more when the capacity of the power storage device is determined.

3.6.3 Power Storage Strategies in Renewable Power Systems

Based on the previous system configuration, power storage strategies can be further explored. The following eight cases have been considered and are shown in Figure 3.40 [86]:

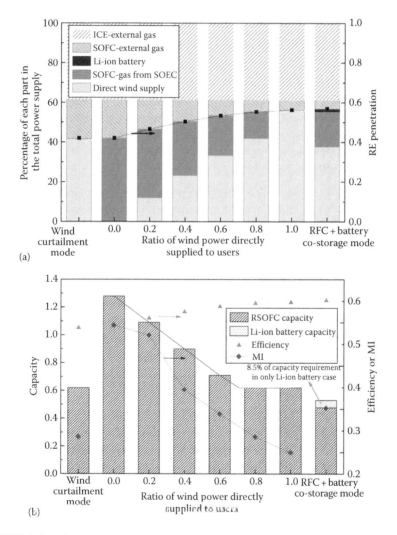

FIGURE 3.40 Comparisons of power storage strategies: (a) supplied power composition and φ_{RE}. (b) MI, system efficiency, and capacity of power storage devices. (Reprinted from *J. Power Sources*, 340, Luo, Y., Shi, Y., Zheng, Y., and Cai, N., Reversible solid oxide fuel cell for natural gas/renewable hybrid power generation systems, 60–70, Copyright 2017, with permission from Elsevier.)

Case No. 1:   Wind curtailment mode. The wind generator is only allowed to supply 40% of its maximum power output to users and the other 60% is curtailed. Remaining power is supplied to users by RSOC and GICE.

Case Nos. 2–7:   A portion of the wind power is sent directly to users and the remaining portion is stored by SOECs. The portion of wind power sent to users varies from 0% to 50% of the maximum wind power, corresponding to Case Nos. 2–7. Remaining power to users is supplied by RSOC and GICE. Here, Case No. 2 is also referred to as PtG mode (all power is converted into gas fuel), where all wind power is electrolyzed; Case No. 7 is the maximum wind penetration mode. Case Nos. 3–6 are referred to as PUPS mode (partial power sent to users and partial power stored).

Case No. 8:   Hybrid storage mode. Here, RSOC and lithium-ion batteries are both used to store power. The power is first stored by RSOC due to its relatively large gas storage capacities, and lithium-ion batteries are used to absorb or supply instantaneous large power supply/demand when wind power output is too high or too low.

Wind curtailment mode (Case No. 1) is currently the most common method to guarantee grid power quality. However, this curtails wind power and limits φ_{RE} and η_{tot}. Case Nos. 2 and 7 are two extreme methods. PtG mode provides the highest power quality and best system dynamic operation stability but requires the largest capacity of power storage and relatively low φ_{RE} and η_{tot}. Maximum wind penetration mode is the opposite case, providing the worst power quality and system dynamic operation stability despite high φ_{RE} and η_{tot}, while reducing power storage capacities. In a distributed system, the optimal strategy should balance efficiencies, renewable power penetrations, system power balances, and capacity requirements instead of going to any one extreme. To balance all these factors, the optimal limited ratio of wind power sent to users should be 0.36. This way, the system can obtain a φ_{RE} of 54.6%, an η_{tot} of 54.2%, and an MI of 0.257 (power unbalance = 6%).

Hybrid power storage systems can further improve system performances in which φ_{RE} and η_{tot} can increase up to 57.1% and 55.2%, respectively [86], while MI can rise from 0.257 to 0.303, corresponding to a decrease in power unbalance from 6% to 4%. The system operation dynamic stability in hybrid power storage systems is also significantly improved, providing satisfactory power qualities to users. Hybrid power storage systems require only 77.5% of the RSOC-only capacities and 8.5% of the battery-only capacities, therefore further reducing the investment costs. The system performance also dramatically improves because all power devices are being more efficiently utilized.

3.7 CHALLENGES AND OUTLOOKS

High-temperature electrolysis using solid oxide electrolysis cells possesses advantages such as high power-to-gas efficiencies and fast reaction rates and provides an alternative, promising technology for energy conversion and storage. In particular, high-temperature electrolysis using solid oxide electrolysis cells provides advantages in the utilization of renewable and nuclear power sources.

Several novel application scenarios for SOECs have been proposed and demonstrated, including high-efficiency power-to-gas production of hydrogen/syngas/methane, reversible operation of SOECs applicable for renewable energy storage, liquid fuel production, biogas upgrading, and complementation with coal plants or coal chemical engineering industries [3]. For power generation using renewable fluctuating and intermittent sources such as photovoltaic plants and wind-driven generators, high-temperature H_2O electrolysis, CO_2 electrolysis, or co-electrolysis of H_2O/CO_2 through SOECs offer promising methods to store and utilize these unstable power sources, even meeting requirements for seasonal power storage.

Nevertheless, SOECs are still in the early stages of development and several challenges must be solved before commercialization [3,95]: (1) insufficient understanding of reaction mechanisms, which requires online, in situ characterization methods; (2) inadequate electrolyte and electrode materials for high performance, low polarization, low cost, and long lifetime operation of SOEC systems; (3) lack of high-performance sealing materials, especially for pressurized operations and scale-up; (4) unoptimized stack assembly and long-term uniform dynamic operations; (5) unoptimized power system integration, design, and control strategies; (6) insufficient auxiliary equipment/component developments, such as effective waste heat recovery and advanced heat exchangers and low-energy F-T synthesis reactors. Therefore, more in-depth research is needed in order to fully develop SOEC technologies.

REFERENCES

1. Stoots C.M., Obrien J.E., Herring J.S., Hartvigsen J.J. 2009. Syngas production via high-temperature coelectrolysis of steam and carbon dioxide. *Journal of Fuel Cell Science & Technology* 6(1): 011014.
2. de Levie R. 1999. The electrolysis of water. *Journal of Electroanalytical Chemistry* 476: 92–93.
3. Shi Y., Luo Y., Li W., Ni M., Cai N. 2015. High temperature electrolysis for hydrogen or syngas production from nuclear or renewable energy. In *Handbook of Clean Energy Systems*, ed. J. Yan. John Wiley & Sons, Ltd., New York.
4. Spacil H.S., Tedmon C.S. 1969. Electrochemical dissociation of water vapor in solid oxide electrolyte cells II. Materials, fabrication, and properties. *Journal of the Electrochemical Society* 116: 1627–1633.
5. Spacil H.S., Tedmon C.S. 1969. Electrochemical dissociation of water vapor in solid oxide electrolyte cells. *Journal of the Electrochemical Society* 116: 1618–1626.
6. Doenitz W., Schmidberger R., Steinheil E. 1980. Hydrogen production by high temperature electrolysis of water vapour. *International Journal of Hydrogen Energy* 5: 55–63.
7. Hauch A., Ebbesen S.D., Jensen S.H., Mogensen M. 2008. Highly efficient high temperature electrolysis. *Journal of Materials Chemistry* 18: 2331.
8. Graves C., Ebbesen S.D., Mogensen M., Lackner K.S. 2011. Sustainable hydrocarbon fuels by recycling CO_2 and H_2O with renewable or nuclear energy. *Renewable and Sustainable Energy Reviews* 15: 1–23.
9. Iacomini C.S. 2004. *Combined Carbon Dioxide/Water Solid Oxide Electrolysis.* The University of Arizona, Tucson, AZ.

10. Stoots C.M. 2006. High-temperature co-electrolysis of H_2O and CO_2 for syngas production. In *2006 Fuel Cell Seminar*. Honolulu, HI.
11. O'Brien J.E., McKellar M.G., Harvego E.A., Stoots C.M. 2010. High-temperature electrolysis for large-scale hydrogen and syngas production from nuclear energy—Summary of system simulation and economic analyses. *International Journal of Hydrogen Energy* 35: 4808–4819.
12. Becker W.L., Braun R.J., Penev M., Melaina M. 2012. Production of Fischer–Tropsch liquid fuels from high temperature solid oxide co-electrolysis units. *Energy* 47: 99–115.
13. Stoots C.M., O'Brien J.E., Condie K.G., Hartvigsen J.J. 2010. High-temperature electrolysis for large-scale hydrogen production from nuclear energy—Experimental investigations. *International Journal of Hydrogen Energy* 35: 4861–4870.
14. Graves C., Ebbesen S.D., Mogensen M. 2011. Co-electrolysis of CO_2 and H_2O in solid oxide cells: Performance and durability. *Solid State Ionics* 192: 398–403.
15. Ebbesen S.D., Høgh J., Nielsen K.A., Nielsen J.U., Mogensen M. 2011. Durable SOC stacks for production of hydrogen and synthesis gas by high temperature electrolysis. *International Journal of Hydrogen Energy* 36: 7363–7373.
16. Jensen S.H., Høgh J.V.T., Barfod R., Mogensen M.B. 2003. High temperature electrolysis of steam and carbon dioxide. In *Proceedings of Risø International Energy Conference*. Roskilde, Denmark.
17. Bierschenk D.M., Wilson J.R., Barnett S.A. 2011. High efficiency electrical energy storage using a methane—Oxygen solid oxide cell. *Energy & Environmental Science* 4: 944–951.
18. Ni M., Leung M.K.H., Leung D.Y.C. 2008. Technological development of hydrogen production by solid oxide electrolyzer cell (SOEC)—Review. *International Journal of Hydrogen Energy* 33(9): 2337–2354.
19. Larminie J., Dicks A. 2000. *Fuel Cell System Explained*. John Wiley & Sons Ltd., New York.
20. Li W., Shi Y., Luo Y., Wang Y., Cai N. 2015. Carbon deposition on patterned nickel/yttria stabilized zirconia electrodes for solid oxide fuel cell/solid oxide electrolysis cell modes. *Journal of Power Sources* 276: 26–31.
21. Gorte R.J., Vohs J.M. 2011. Catalysis in solid oxide fuel cells. *Annual Review of Chemical and Biomolecular Engineering* 2: 9–30.
22. Li C., Shi Y., Cai N. 2013. Carbon deposition on nickel cermet anodes of solid oxide fuel cells operating on carbon monoxide fuel. *Journal of Power Sources* 225: 1–8.
23. Millichamp J. et al. 2013. A study of carbon deposition on solid oxide fuel cell anodes using electrochemical impedance spectroscopy in combination with a high temperature crystal microbalance. *Journal of Power Sources* 235: 14–19.
24. Koh J., Yoo Y., Park J., Lim H.C. 2002. Carbon deposition and cell performance of Ni-YSZ anode support SOFC with methane fuel. *Solid State Ionics* 149: 157–166.
25. O'Hayre R.P., Cha S.W., Colella W., Prinz F.B. 2008. *Fuel Cell Fundamentals*. John Wiley & Sons, Ltd., New York.
26. Huang K., Goodenough J.B. 2009. *Solid Oxide Fuel Cell Technology: Principles, Performance and Operations*. Woodhead Publishing, Oxford, U.K.
27. EG&G Technical Services. 2005. *Fuel Cell Handbook*. Contract No. DE-AM26-99FT40575, U.S. Department of Energy, Office of Fossil Energy, National Energy Technology Laboratory, WV.
28. O'Hayre R.P., Cha S.W., Colella W., Prinz F.B. 2006. *Fuel Cell Fundamentals*. John Willey & Sons, Ltd., Hoboken, NJ.
29. Kee R.J., Zhu H.Y., Sukeshini A.M., Jackson G.S. 2008. Solid oxide fuel cells: operating principles, current challenges, and the role of syngas. *Combustion Science and Technology* 180: 1207–1244.

30. Gorte R.J., Kim H., Vohs J.M. 2002. Novel SOFC anode for the direct electrochemical oxidation of hydrocarbon. *Journal of Power Sources* 106: 10–15.

31. Bieberle A., Meier L.P., Gauckler L.J. 2001. The electrochemistry of Ni pattern anodes used as solid oxide fuel cell model electrodes. *Journal of the Electrochemical Society* 148: A646–A656.

32. Rao M.V., Fleig J., Zinkevich M., Aldinger F. 2010. The influence of the solid electrolyte on the impedance of hydrogen oxidation at patterned Ni electrodes. *Solid State Ionics* 181: 1170–1177.

33. Goodwin D.G., Zhu H., Colclasure A.M., Kee R.J. 2009. Modeling electrochemical oxidation of hydrogen on Ni—YSZ pattern anodes. *Journal of the Electrochemical Society* 156: B1004–B1021.

34. Vogler M., Bieberle A., Gauckler L., Warnatz J., Bessler W.G. 2009. Modelling study of surface reactions, diffusion, and spillover at a Ni/YSZ patterned anode. *Journal of the Electrochemical Society* 156: B663–B672.

35. Shi Y., Cai N., Mao Z. 2012. Simulation of EIS spectra and polarization curves based on Ni/YSZ patterned anode elementary reaction models. *International Journal of Hydrogen Energy* 37: 1037–1043.

36. Utz A., Störmer H., Gerthsen D., Weber A., Ivers-Tiffée E. 2011. Microstructure stability studies of Ni patterned anodes for SOFC. *Solid State Ionics* 192: 565–570.

37. Bessler W.G. et al. 2010. Model anodes and anode models for understanding the mechanism of hydrogen oxidation in solid oxide fuel cells. *Physical Chemistry Chemical Physics* 12: 13888.

38. Ehn A. et al. 2010. Electrochemical investigation of nickel pattern electrodes in $H_2\tilde{O}H_2O$ and $CO\tilde{O}CO_2$ atmospheres. *Journal of the Electrochemical Society* 157: B1588–B1596.

39. Utz A., Störmer H., Leonide A., Weber A., Ivers-Tiffée E. 2010. Degradation and relaxation effects of Ni patterned anodes in H_2—H_2O atmosphere. *Journal of the Electrochemical Society* 157: B920–B930.

40. Utz A., Leonide A., Weber A., Ivers-Tiffée E. 2011. Studying the CO–CO_2 characteristics of SOFC anodes by means of patterned Ni anodes. *Journal of Power Sources* 196: 7217–7224.

41. Utz A., Hansen K.V., Norrman K., Ivers-Tiffée E., Mogensen M. 2011. Impurity features in Ni-YSZ-H_2-H_2O electrodes. *Solid State Ionics* 183: 60–70.

42. Li W., Shi Y., Luo Y., Wang Y., Cai N. 2016. Carbon monoxide/carbon dioxide electrochemical conversion on patterned nickel electrodes operating in fuel cell and electrolysis cell modes. *International Journal of Hydrogen Energy* 41: 3762–3773.

43. Alzate-Restrepo V., Hill J.M. 2008. Effect of anodic polarization on carbon deposition on Ni/YSZ anodes exposed to methane. *Applied Catalysis A: General* 342: 49–55.

44. Lin Y., Zhan Z., Liu J., Barnett S. 2005. Direct operation of solid oxide fuel cells with methane fuel. *Solid State Ionics* 176: 1827–1835.

45. Shi Y. et al. 2013. Experimental characterization and modeling of the electrochemical reduction of CO_2 in solid oxide electrolysis cells. *Electrochimica Acta* 88: 644–653.

46. Janardhanan V.M., Deutschmann O. 2006. CFD analysis of a solid oxide fuel cell with internal reforming: Coupled interactions of transport, heterogeneous catalysis and electrochemical processes. *Journal of Power Sources* 162: 1192–1202.

47. Yoshinaga M. et al. 2011. Deposited carbon distributions on nickel film/oxide substrate systems. *Solid State Ionics* 192: 571–575.

48. Williford R.E., Chick L.A., Maupin G.D., Simner S.P., Stevenson J.W. 2003. Diffusion limitations in the porous anodes of SOFCs. *Journal of the Electrochemical Society* 150: A1067–A1072.

49. Burghaus U. 2014. Surface chemistry of CO_2—Adsorption of carbon dioxide on clean surfaces at ultrahigh vacuum. *Progress in Surface Science* 89: 161–217.

50. Bartos B., Freund H.J., Kuhlenbeck H., Neumann M., Lindner H., Müller K. 1987. Adsorption and reaction of CO_2 and CO_2/O CO-adsorption on Ni(110): Angle resolved photoemission (ARUPS) and electron energy loss (HREELS) studies. *Surface Science* 179: 59–89.
51. Hayek H., Kramer R., Paál Z. 1997. Metal-support boundary sites in catalysis. *Applied Catalysis A: General* 162: 1–15.
52. Ali A., Wen X., Nandakumar K., Luo J., Chuang K.T. 2008. Geometrical modeling of microstructure of solid oxide fuel cell composite electrodes. *Journal of Power Sources* 185: 961–966.
53. Li W., Wang H., Shi Y., Cai N. 2013. Performance and methane production characteristics of H_2O–CO_2 co-electrolysis in solid oxide electrolysis cells. *International Journal of Hydrogen Energy* 38: 11104–11109.
54. Stoots C., O'Brien J., Hartvigsen J. 2009. Results of recent high temperature coelectrolysis studies at the Idaho National Laboratory. *International Journal of Hydrogen Energy* 34: 4208–4215.
55. Kim-Lohsoontorn P., Bae J. 2011. Electrochemical performance of solid oxide electrolysis cell electrodes under high-temperature coelectrolysis of steam and carbon dioxide. *Journal of Power Sources* 196: 7161–7168.
56. Vernoux P., Guindet J., Kleitz M. 1998. Gradual internal methane reforming in intermediate-temperature solid-oxide fuel cells. *Journal of the Electrochemical Society* 145: 3487–3492.
57. Takeguchi T., Kikuchi R., Yano T., Eguchi K., Murata K. 2003. Effect of precious metal addition to Ni-YSZ cermet on reforming of CH4 and electrochemical activity as SOFC anode. *Catalysis Today* 84: 217–222.
58. Li W., Shi Y., Luo Y., Cai N. 2013. Elementary reaction modeling of CO_2/H_2O co-electrolysis cell considering effects of cathode thickness. *Journal of Power Sources* 243: 118–130.
59. Sun C., Stimming U. 2007. Recent anode advances in solid oxide fuel cells. *Journal of Power Sources* 171: 247–260.
60. Li W., Shi Y., Luo Y., Cai N. 2014. Theoretical modeling of air electrode operating in SOFC mode and SOEC mode: The effects of microstructure and thickness. *International Journal of Hydrogen Energy* 39: 13738–13750.
61. Li W., Shi Y., Luo Y., Cai N. 2015. Elementary reaction modeling of solid oxide electrolysis cells: Main zones for heterogeneous chemical/electrochemical reactions. *Journal of Power Sources* 273: 1–13.
62. Li C., Shi Y., Cai N. 2010. Elementary reaction kinetic model of an anode-supported solid oxide fuel cell fueled with syngas. *Journal of Power Sources* 195: 2266 2282.
63. Bertei A., Nucci B., Nicolella C. 2013. Microstructural modeling for prediction of transport properties and electrochemical performance in SOFC composite electrodes. *Chemical Engineering Science* 101: 175–190.
64. Luo Y., Shi Y., Li W., Ni M., Cai N. 2014. Elementary reaction modeling and experimental characterization of solid oxide fuel-assisted steam electrolysis cells. *International Journal of Hydrogen Energy* 39: 10359–10373.
65. Wu Y., Shi Y., Luo Y., Cai N. 2016. Elementary reaction modeling and experimental characterization of solid oxide direct carbon-assisted steam electrolysis cells. *Solid State Ionics* 295: 78–89.
66. Zhang W., Yu B., Chen J., Xu J. 2008. Hydrogen production through solid oxide electrolysis at elevated temperatures. *Progress in Chemistry*: 778–787.
67. Tao G., Butler B., Virkar A.V. 2011. Hydrogen and power by fuel-assisted electrolysis using solid oxide fuel cells. *ECS Transactions* 1: 2929–2939.
68. Wang W., Vohs J.M., Gorte R.J. 2007. Hydrogen production Via CH_4 and CO assisted steam electrolysis. *Topics in Catalysis* 46: 380–385.

69. Xu H., Chen B., Ni M. 2016. Modeling of direct carbon-assisted solid oxide electrolysis ell (SOEC) for syngas production at two different electrodes. *Journal of the Electrochemical Society* 163(11): F3029–F3035.
70. Wang Y., Liu T., Fang S., Xiao G., Wang H., Chen F. 2015. A novel clean and effective syngas production system based on partial oxidation of methane assisted solid oxide co-electrolysis process. *Journal of Power Sources* 277: 261–267.
71. Shao L., Wang S., Qian J., Ye X., Wen T. 2013. Optimization of the electrode-supported tubular solid oxide cells for application on fuel cell and steam electrolysis. *International Journal of Hydrogen Energy* 38: 4272–4280.
72. Luo Y. et al. 2015. Experimental characterization and theoretical modeling of methane production by H_2O/CO_2 co-electrolysis in a tubular solid oxide electrolysis cell. *Journal of the Electrochemical Society* 162: F1129–F1134.
73. Luo Y., Shi Y., Li W., Cai N. 2014. Comprehensive modeling of tubular solid oxide electrolysis cell for co-electrolysis of steam and carbon dioxide. *Energy* 70: 420–434.
74. Luo Y., Shi Y., Li W., Cai N. 2015. Dynamic electro-thermal modeling of co-electrolysis of steam and carbon dioxide in a tubular solid oxide electrolysis cell. *Energy* 89: 637–647.
75. Stempien J.P., Sun Q., Chan S.H. 2013. Performance of power generation extension system based on solid-oxide electrolyzer cells under various design conditions. *Energy* 55: 647–657.
76. Fu Q., Mabilat C., Zahid M., Brisse A., Gautier L. 2010. Syngas production via high-temperature steam/CO_2 co-electrolysis: an economic assessment. *Energy & Environmental Science* 3: 1382–1397.
77. Wang M., Wang Z., Gong X., Guo Z. 2014. The intensification technologies to water electrolysis for hydrogen production – A review. *Renewable and Sustainable Energy Reviews* 29: 573–588.
78. O Brien, J.E., Stoots C., Herring J.S., Hartvigsen J.J. 2005. High-temperature electrolysis for hydrogen production from nuclear energy. In *11th International Topical Meeting on Nuclear Reactor Thermal-Hydraulics (NURETH-11)*. Avignon, France.
79. Steinberger-Wilkens R., Lehnert W. 2010. *Innovations in Fuel Cell Technologies*. RSC Publishing, Cambridge, U.K.
80. Ridjan I., Mathiesen B.V., Connolly D., Dui N. 2013. The feasibility of synthetic fuels in renewable energy systems. *Energy* 57: 76–84.
81. Simith I. 1999. *CO_2 Reduction Prospects for Coal*. ISBN 92-9029-336-5: IEA research report, London, UK.
82. Unruh D., Pabst K., Schaub G. 2010. Fischer–Tropsch Synfuels from Biomass: Maximizing carbon efficiency and hydrocarbon yield. *Energy & Fuels* 24: 2634–2641.
83. Mills G.A., Steffgen F.W. 1974. Catalytic methanation. *Catalysis Reviews* 8: 159–210.
84. Dong Z., Zhao J., Wen F., Xue Y. 2014. From smart grid to energy internet: basic concept and research. *Automation of Electric Power Systems* 38: 1–11.
85. Jensen S.H. et al. 2015. Large-scale electricity storage utilizing reversible solid oxide cells combined with underground storage of CO_2 and CH_4. *Energy Environment Science* 8: 2471–2479.
86. Luo Y., Shi Y., Zheng Y., Cai N. 2017. Reversible solid oxide fuel cell for natural gas/renewable hybrid power generation systems. *Journal of Power Sources* 340: 60–70.
87. Fraser A.M., Swinney H.L. 1986. Independent coordinates for strange attractors from mutual information. *Physical Review A* 33: 1134–1140.
88. Maes F., Collignon A., Vandermeulen D., Marchal G., Suetens P. 1997. Multimodality image registration by maximization of mutual information. *IEEE Transactions on Medical Imaging* 16: 187–198.
89. Viola P., Wells W.M. 1997. Alignment by maximization of mutual information. *International Journal of Computer Vision* 24: 137–154.

90. Peng H., Long F., Ding C. 2005. Feature selection based on mutual information: Criteria of max-dependency, max-relevance, and min-redundancy. *IEEE Transactions on Pattern Analysis and Machine Intelligence* 27: 1226–1238.
91. Steuer R., Kurths J., Daub C.O., Weise J., Selbig J. 2002. The mutual information: Detecting and evaluating dependencies between variables. *Bioinformatics* 18: S231–S240.
92. Shannon C.E. 1948. The mathematical theory of communication. *The Bell System Technical Journal* 27(379–424): 623–656.
93. Kolmogorov A. 1968. Logical basis for information theory and probability theory. *IEEE Transactions on Information Theory* 14: 662–664.
94. Cover T.M., Thomas J.A. 1991. *Elements of Information Theory*. Wiley, New York.
95. Ebbesen S.D., Jensen S.H., Hauch A., Mogensen M.B. 2014. High temperature electrolysis in alkaline cells, solid proton conducting cells, and solid oxide ells. *Chemical Reviews* 114: 10697–10734.

4 Flame Fuel Cells

4.1 INTRODUCTION

Distributed energy systems are becoming increasingly important in energy markets, especially for small-scale applications. Of these, fuel cell systems are promising due to their energy efficiency, low/zero emission, and low noise level, and co-generation of heat and power (CHP) [1–4].

Fuel cells are efficient and noiseless devices that can directly convert chemical energy into electricity without limitations of the Carnot cycle [5] and have attracted increasing attention in recent years. Fuel cells can be classified into alkaline fuel cells (AFCs), phosphoric acid fuel cells (PAFCs), molten carbonate fuel cells (MCFCs), proton-exchange membrane fuel cells (PEMFCs), solid oxide fuel cells (SOFCs), and direct liquid fuel fuel cells (DLFFCs). By combining an SOFC with a micro-CHP to form an energy system, high energy efficiency and fuel flexibility can be achieved [6–12]. An example of this is a micro-CHP system with SOFC, produced by ENE-FARM-S in Japan that can achieve electrical efficiencies of 46.5% [12].

Scientifically, the driving force for the conversion of chemical energy to electricity in SOFCs is the chemical-potential gradient from fuel to oxidant. Conventional SOFCs operated in a dual-chamber manner, where fuel in the anode chamber and air in the cathode chamber are isolated from each other. Due to differences in thermal expansion between the ceramic cell components however, sealing remains a serious challenge in dual-chamber SOFCs, especially at high temperatures. To solve this problem, two sealant-free SOFC configurations have been proposed and studied recently. One sealant-free SOFC is a single-chambered SOFC [13–16], in which the anode and cathode are exposed to the same fuel–air mixture in a single gas chamber [13,17]. The principle of this single-chamber SOFC is based on the different catalytic activities and selectivity of the two electrodes toward the fuel–air mixture [18,19] and presents great challenges to the functionality of the catalysts. The other sealant-free SOFC configuration is the flame fuel cell (FFC) that combines a flame with an SOFC in a "no-chamber" setup. Here, the anode directly exposes to a fuel-rich flame, while the cathode is exposed to air [20], eliminating the need for a sealant. Compared to conventional dual-chamber SOFCs and the single-chamber SOFCs, FFCs possess several advantages, including fuel flexibility and sealing free, which lead to simple setup and rapid start-up [21].

FFCs were first proposed by Horiuchi et al. [21] from Shinko Electric Industries in Japan. After the concept was proposed in 2004, research institutes and universities including the Toyohashi University of Technology [22,23], Shinko Electric Industries in Japan [24–27], the University of Duisburg-Essen [24,25], the University of Heidelberg in Germany [25], Georgia Institute of Technology, Syracuse University in the United States [28–35], Institute Carnot CIRIMAT UMR CNRS in France [36], Nanjing University of Technology [17,37], Jinning Institute [38], Harbin Institute of

Technology [20,39], Tsinghua University [40–47], and the University of Strathclyde in the United Kingdom [48] have conducted extensive research on FFCs.

This "no-chamber" FFC setup combines a fuel-rich flame with an SOFC, in which the fuel-rich flame provides both heat and fuels for the SOFC. Suitable operating temperatures (usually 923~1073 K) are provided by the reaction heat from the combustion of the fuel-rich flame without the need for additional thermal management systems. The fuel-rich flame also serves as a fuel reformer for hydrocarbons, consuming oxygen at the anode and maintaining the difference of oxygen concentration between the two electrodes. In addition to producing electricity, FFCs can produce large amounts of high-temperature heat, demonstrating great potential in distributed generation systems and CHP systems.

Compared with other types of fuel cells, FFCs has the following advantages:

1. *Fuel flexibility*: The fuel-rich flame can be used as a reformer for SOFCs, producing syngases such as H_2 and CO from the original fuels. Because of this, FFCs can tolerate almost any type of fuels, including gases, liquids, or solids.
2. *Simple structure*: The devices and construction of FFCs are simple, avoiding common problems of sealing at high temperatures. The fuel-rich combustion can also replace start-up burners and provide heat for heat-consuming devices such as evaporators, reformers, and air preheaters [41].
3. *High combined efficiency*: Although the power-generating efficiencies of FFCs are still low, they can achieve relatively high comprehensive efficiencies when combining heating and power for the cascading utilization of fuel chemical energies.
4. *Rapid start-up*: The fuel-rich flame can provide a heat source for SOFCs to start up without the need for any additional heat management systems or electric starting equipment, allowing easier and faster system start-up. FFCs can be directly started by the fuel-rich flame within seconds in given conditions.
5. *Low cost*: Compared to single-chamber setups, the cathode and anode of FFCs are separately located in the oxidant gases and the flame, therefore, the highly selective catalysts are not required. This can potentially lower fuel cell costs.

Despite these advantages, FFC technology is far from commercialization and requires extensive research.

4.2   WORKING PRINCIPLES AND EFFICIENCY ANALYSIS

An FFC unit mainly consists of a fuel-rich combustion burner and an SOFC, as shown in Figure 4.1. The SOFC consists of a cathode, electrolyte, and anode. The fuel-rich flame at the anode of the SOFC consumes oxygen-producing CO and H_2, which are ideal fuels for the SOFC. When the SOFC is working, the produced CO and H_2 is oxidized by oxygen ions, producing electrons at the anode. At the cathode, oxygen in air is reduced and oxygen ions are formed. The chemical potential gradient between the

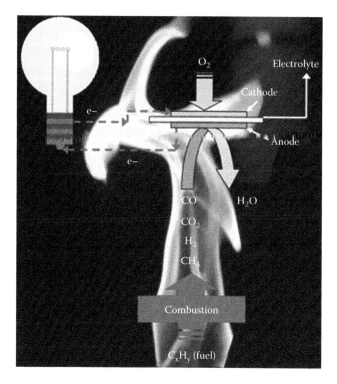

FIGURE 4.1 Schematic diagram of FFC working principles [5]. (Reprinted from *Prog. Energy Combust. Sci.*, 37, Walther, D.C. and Ahn, J., Advances and challenges in the development of power-generation systems at small scales, 583–610. Copyright 2011, with permission from Elsevier.)

anode and the cathode is maintained by the fuel-rich flame, which consumes oxygen at the anode and forms a different gas environment without the need for any sealing. The heat generated by combustion can also be used to start up the SOFC and maintain suitable temperature environments.

The working processes of FFCs can be divided into the following three parts to analyze its efficiency:

1. *Partial oxidation reforming*: Fuel-rich combustion occurs in the burner, producing syngas through partial oxidation reforming:

$$CH_4 + \frac{2}{\phi}(O_2 + 3.76N_2) \rightarrow aCO_2 + bCO + cH_2O + dH_2 + \frac{2 \times 3.76}{\phi}N_2 \quad (4.1)$$

This reforming process consists of the following two reactions:

$$CH_4 + 2O_2 \rightarrow CO_2 + 2H_2O \quad (4.2)$$

$$CH_4 + 1/2O_2 \rightarrow CO + 2H_2 \quad (4.3)$$

Equation 4.2 is the combustion of CH_4, and Equation 4.3 is the partial oxidation of CH_4. Simply put, the original fuels (such as methane) are partly reformed into syngases (CO and H_2, which are the available fuels for SOFCs) with the other chemical energies of the original fuels being converted into sensible heat. The reforming efficiency is defined as the ratio of the heat value of the produced H_2 and CO to the heat value of the supplied fuel, referred to as the combustion efficiency, given by [25,45]

$$\varepsilon_{re} = \frac{\text{Heat value of } H_2 \text{ and CO}}{\text{Heat value of supplied fuel}} \tag{4.4}$$

This equation can be simplified as Equation 4.5:

$$\varepsilon_{re} = 1 - 1/\phi \tag{4.5}$$

2. *Fuel utilization*: The syngases reformed by the partial oxidation of the original fuels are partly utilized by the anode of the SOFC. Because gas combustion kinetics is much faster than electrochemical kinetics, most of the syngases will leave without being used [33].

In this part, the fuel utilization efficiency is defined by Equation 4.6:

$$\varepsilon_{fu} = \frac{H_2 \text{ and CO consumed by SOFC}}{\text{Total } H_2 \text{ and CO after flame}} \tag{4.6}$$

The fuel utilization efficiency can be calculated by [25,42,45]

$$\varepsilon_{fu} = \frac{i}{\varepsilon_{re} 4 F f_{stoich} V_{fuel} / V^M} \tag{4.7}$$

where
i (A) is the current passing through the SOFC
F is Faraday's constant
f_{stoich} is the stoichiometric oxygen/fuel ratio
V_{fuel} (m^3/s) is the inflow volume velocity of the cold fuel
V^M (m^3/mol) is the molar volume at standard conditions, which is 22.4×10^{-3}

3. *Electrochemical reactions*: The fuels are oxidized by oxygen ions and produce electrons at the anode of the SOFC:

$$H_2 + O^{2-} \rightarrow H_2O + 2e^- \tag{4.8}$$

$$CO + O^{2-} \rightarrow CO_2 + 2e^- \tag{4.9}$$

At the cathode of the SOFC, the oxygen in air is reduced by electrons and produces oxygen ions:

$$O_2 + 4e^- \rightarrow 2O^{2-} \tag{4.10}$$

The electrochemical reactions can only occur at the three-phase boundary in the electrode where the electronic conductor, ionic conductor, and gas phase come together. The SOFC electrical efficiency ε_{el} of this portion is defined by Equation 4.11:

$$\varepsilon_{el} = \frac{\text{Electrical power output}}{\text{Heat value of } H_2 \text{ and CO cunsumed by SOFC}} \tag{4.11}$$

Based on combustion efficiency, fuel utilization efficiency, and SOFC electrical efficiency, the total electrical efficiency (ε) of an SOFC can be given as [25,45]

$$\varepsilon = \frac{\text{Electrical power output}}{\text{Heat value of supplied fuel}} = \varepsilon_{re}\varepsilon_{fu}\varepsilon_{el} \tag{4.12}$$

Therefore, in order to improve the efficiency of FFCs, the combustion efficiency, fuel utilization efficiency, and SOFC electrical efficiency must be improved. Combustion efficiencies are dominated by burner structures and operation parameters such as gas velocities and equivalence ratios. Utilization efficiencies are determined by reaction residence times, which can be increased through the proper matching of cell numbers in SOFCs and reactant gas flow rates. SOFC electrical efficiencies are affected by burner gas compositions, temperatures, and material activities of electrodes and electrolytes. In addition, the reactivity at the three-phase boundary is also an important issue.

Vogler et al. [25] studied the efficiencies of a no-chamber, direct-flame SOFC with methane/air-rich flames provided by a flat-flame burner. The experimental results showed that the highest efficiency of the tested FFC was only 0.45% when the flame speed was 30 cm/s and the fuel/air ratio was 1.1.

Research related to the electrical efficiencies of FFCs [32–34,41,42,45,47] show that low fuel utilization efficiency is the major cause for low total efficiency. Due to relatively quick combustion kinetics and slow electrochemical kinetics, reducing gas flow rates can be a potential method to increase fuel utilization efficiencies.

4.3   FUEL-RICH COMBUSTION

Fuel-rich combustion burners are one of the most important components of FFCs because the uniformity and stability of the flame are dominating factors for proper operation of the cell. Operational stability, FFC lifetime, and electrical efficiency all closely related with combustion status and combustion products. Therefore, providing ideal flames for SOFCs is the key to the application of FFC systems.

For combustion efficiencies, increasing equivalence ratios is a promising way to promote the conversion of original fuels to syngas. With increasing equivalence ratios, higher syngas production is expected for higher concentrations of reactants at the anode of the SOFC. However, when equivalence ratios are too high, the stability of the rich combustion is lowered. When equivalence ratios are higher than the critical value, combustion temperatures become too low to hold a stable flame and undesired phenomenon such as "blow-off" might occur.

Stable and suitable temperature environments for FFCs are essential when increasing equivalence ratios to improve combustion efficiency, avoiding damages caused by fluctuations in temperature near the SOFC. The stability of fuel-rich combustion under higher equivalence ratios is important because the heat produced by the combustion is required to keep the SOFC in a suitable temperature environment.

4.3.1 TYPES OF BURNERS

FFCs based on bunsen burners were originally used by Horiuchi et al. [21] to combust n-butane, kerosene, paraffin wax, and wood, proving that electricity can be produced from various types of fuels in FFCs. However, the typical cone flame from Bunsen burners is not well distributed, causing significant variations in temperature and gas composition along the anode of SOFCs. The thermal stress caused by this large temperature gradient can cause considerable damage to SOFCs, resulting in unacceptable thermal shock.

Alcohol burners [17,37], household gas cookers [20,39], and quartz tubes [30] were also explored for the partial oxidation reforming of original fuels, but the same issues with flame structure and stability remains to be solved.

To solve the problem of flame uniformity, McKenna flat-flame burners were employed by the FFCs, which was proposed by Kronemayer et al. to replace Bunsen burners [24]. Their proposed burner provided steady and uniform flames in their studies. Similarly, Wang et al. [43–45] used a Hencken burner with multielement diffusion flames to obtain flame uniformity in both vertical and horizontal directions, and also studied its performance when integrated with a button cell, a planar SOFC, and a microtubular SOFC.

The gas composition of combustion exhausts can also significantly influence FFC performances. The stability of rich combustion is usually limited by a maximum equivalence ratio, called the fuel-rich flammability limit, because equivalence ratios higher than this will cause the flame to be unstable. In order to expand the operation region of the fuel rich burner, Wang et al. [41,42] used porous media burners with advantageous combustion characteristics and studied FFC performance through its integration with microtubular SOFCs.

The structures and characteristics of McKenna flat-flame burners, Hencken burners, and porous media burners used in FFCs are reviewed in the following sections.

4.3.1.1 McKenna Burner

Premixed flames from McKenna burners [49,50] are considered to be a one-dimensional uniform flat flame and are usually used as a standard burner for the calibration of optical measuring instruments [51–53]. Researchers have also used it to study the mechanisms of soot formation.

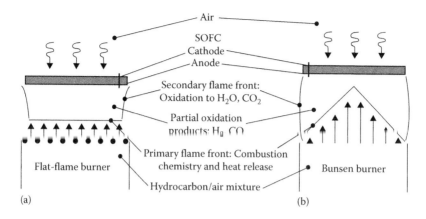

FIGURE 4.2 Operation principles of an FFC with McKenna burner (a) and Bunsen burner (b) [24]. (Reprinted from *J. Power Sources*, 166, Kronemayer, H., Barzan, D., Horiuchi, M., Suganuma, S., Tokutake, Y. et al., A direct-flame solid oxide fuel cell (DFFC) operated on methane, propane, and butane, 120–126. Copyright 2007, with permission from Elsevier.)

In a McKenna burner, a mixture of fuel and air is fed into a porous plug made up of stainless steel or copper at the bottom, allowing for a well-distributed flame. Migliorini et al. [54] studied the fuel-rich combustion of ethylene using a McKenna burner with a stainless steel porous plug and discovered that the amount of soot varied along the radial direction, suggesting a nonuniform flame. On the other hand, they also studied a McKenna burner using a copper porous plug and found that it produced a well-distributed flat flame.

Kronemayer et al. [24] built a McKenna flat-flame burner to replace the Bunsen burner in an FFC, as shown in Figure 4.2. Some studies on FFCs were conducted upon McKenna burners, which would produce stable and well-distributed flat flames, and uniform temperature fields when the FFC is working.

Although stable and well-distributed flat flames can be achieved using McKenna burners, there are serious challenges for FFC applications. The rich flammable limits of McKenna burners are relatively low for the production of syngases. For partial oxidation reforming, higher equivalence ratios are needed so that more H_2 and CO can be produced, allowing for higher combustion efficiencies. Hence, a combustion burner with larger flammable limits is necessary for further developments in FFC.

4.3.1.2 Hencken Burner

Hencken burners are used in particle combustion and coal pyrolysis experiments [55–61] and produce a nonpremixed, multielement diffusion flat flame. Wooldridge et al. [62] studied the synthesis of nano-SiO_2 on a Hencken burner and demonstrated that it can achieve a roughly one-dimensional diffusion flame with a relatively wide adjusting range for experimental variables. Wang et al. [43–45] employed a Hencken burner in an FFC configuration, taking advantage of its flame uniformity in both vertical and horizontal directions.

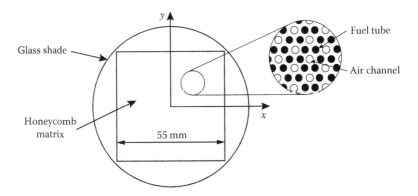

FIGURE 4.3 Top view of the Hencken burner used by Wang et al. (Reproduced from Wang, Y.Q. et al., *J. Electrochem. Soc.*, 160, F1241. Copyright 2013, The Electrochemical Society.)

A Hencken burner consists of arrays of capillary tube that are inserted into a honeycomb matrix, producing a uniform flat diffusion flame. Fuels can only flow through the individually sealed tubes, and oxidants flow around the honeycomb matrix. It could also be operated reversely, oxidants can flow through the tubes while fuels flow through the surrounding channels. With this unique arrangement, fuels and oxidants can only be mixed at the outside of the burner, avoiding flashbacks caused by premixing. Because Hencken burners consist of many small flame bunches, the flames are considered flat and uniform at a macro level.

The top view of the Hencken burner used by Wang et al. [44] is shown in Figure 4.3. The outlet of the burner has a square dimension of 55 mm × 55 mm, with many tubes being evenly inserted in the honeycomb matrix. The Hencken burner used is made up of an array of about 232 capillary tubes, with an inner diameter of 1.2 mm and an outer diameter of 1.5 mm. A round glass shade is added to the outside of the burner to prevent disturbances from the outside.

Figure 4.4 shows the transverse temperature profiles at different heights above the burner surface (a) and the axial temperature profiles for different equivalence ratios (b) measured by Wang et al. [44]. The experimental results showed that Hencken burners can provide steady and uniform temperature environments for SOFC operations, both in vertical and horizontal directions.

4.3.1.3 Porous Combustors

Porous combustors refer to the addition of porous solid mediums such as ceramic foam or alumina pellets to enhance combustion stabilities. Due to the excellent heat conduction and radiation of porous medium materials, reaction heat produced by combustion can be transferred to the upstream through the porous media, preheating premixed reactants, and expanding flammability limits. Using porous media, maximum temperatures during combustion can be higher than the adiabatic flame temperatures of the premixed gases, allowing super-adiabatic combustion [63].

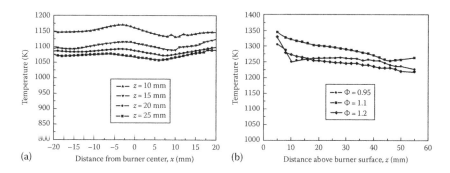

FIGURE 4.4 Transverse (a) and axial (b) temperature profiles of a Hencken burner. (Reproduced from Wang, Y.Q. et al., *J. Electrochem. Soc.*, 160, F1241, Copyright 2013, The Electrochemical Society.)

Porous combustors are already widely used for the combustion of low heating value gases in ultralean conditions or for hydrogen production in fuel-rich combustion [64–67].

Because the flame front is propagating and stablized in the combustion zone, both filtration and stable combustion are explored [68]. During filtration combustion, the flame front propagates at a filtration velocity along the axial direction of the burner. As the flame propagates downstream, the inlet premixed unburned gases can be effectively preheated by the porous media because the flame front passes through these mediums in the previous stage. This feature can significantly expand flammability limits and increase equivalence ratios in fuel-rich conditions, further increasing reforming efficiencies of the original fuel [69]. Most experiments and simulations of hydrogen production in porous media burners were conducted based on filtration combustion [64,70–72]. Dhamrat and Ellzey [73] studied syngas production using filtration combustion in a porous media burner. Their experiment showed that reforming efficiencies of up to 73% can be obtained at an equivalence ratio of 2.5 when fueled with methane. However, because the flame front is always propagating along the burner, the temperature environment in the filtration combustion process is not stable. Large fluctuations in temperature and thermal cycles can easily result in failure of SOFCs.

Stable combustion in porous combustors is desirable to obtain steady and suitable temperature for SOFCs. To ensure stable operations, stable combustion is considered to be a safe and feasible method for the system reliability. Using two layers of porous-medium with different physical properties, such as different porosities, a stable flame between the upstream porous zone and downstream zone can be achieved in certain conditions. Several techniques have been applied to achieve stabilized flames in porous media combustors, with velocity stabilization and Peclet number stabilization being the most frequently used. Velocity stabilization usually occurs in conditions where the flame velocity is equal to the local gas velocity. Here, any disturbances from ambience could interrupt the flame stabilization and the flame position will change.

The Peclet number is defined as $Pe = (S_L * d_m)/\alpha$, where S_L is the laminar flame velocity, d_m is the equivalent pore diameter of the porous media, and α is the thermal diffusivity of the gas mixture [42]. There is a critical Peclet number Pe_c for each type of fuel, and the concept of two-layer porous media combustion is based on this critical number. In the upstream porous layer, a smaller pore diameter is required to lower the Peclet number smaller than the Pe_c so that the flame cannot propagate into the upstream layer. In the downstream layer, a larger pore diameter is required making the Peclet number larger than the Pe_c. By achieving this, a stable flame is achieved at the interface between the two layers. Previous studies on methane combustion in two-layer porous media burners mainly focused on fuel-lean combustion, with little focus on fuel-rich combustion. Pedersen-Mjaanes et al. [74] studied syngas production using methane as a fuel in a two-layer porous media combustor and achieved a reforming efficiency of 45% when the equivalence ratio was 1.85. Al-Hamamre et al. [75] suggested that fuel-rich combustion is a potential technique for fuel reforming in SOFC systems.

Wang et al. [41,42] studied a microtubular FFC system built based on a two-layered porous combustor consisting of two types of alumina pellets with different diameters, using methane as the fuel. Figure 4.5 shows the schematic of the FFC reactor with the porous combustor. The combustion chamber for the porous media burner is 54 mm in inner diameter and 200 mm in length. The burner is inserted inside an insulation cylinder and surrounded by a ceramic insulation layer outside. A two-layer porous structure is used to produce a stable rich flame, with a 20 mm long upstream layer of 2–3 mm Al_2O_3 beads and a 60 mm long downstream layer of 5 mm Al_2O_3 beads. The flame stability limit of the porous combustor used by Wang et al. [42] is shown in Figure 4.6. It can be seen that by increasing the equivalence ratio from 1.4 to 1.8, the stable condition inlet velocity slows. When the equivalence ratio exceeds 1.8, the flame becomes unstable and propagates downstream. When the equivalence ratio is fixed at 1.6, the temperature profile along the burner and the gas composition for various inlet gas velocities are shown in Figure 4.7. The volumetric heat release rates increase with increasing gas velocities, causing both the maximum temperature of the flame and the temperature of the exhaust gas to increase. The compositions of the product gas are determined by both combustion temperatures and the residence time. Therefore, the burner efficiency increases at first and then decreases with increasing gas velocities. For the same gas velocity, temperature distributions and compositions at downstream for various equivalence ratios are shown in Figure 4.8. By increasing equivalence ratios, the mole fractions of H_2 and CO are increased, leading to burner efficiency increases from 33.1% to 41.1%.

4.4 FFC PERFORMANCE

Horiuchi et al. [21] from the Core technology research laboratory of Japan are pioneers in proposing the concept of FFCs, directly placing an SOFC onto a free flame in a no-chamber configuration where combustible gases, liquids, and solids including n-butane, kerosene, paraffin wax, and wood can be used and integrated with SOFCs.

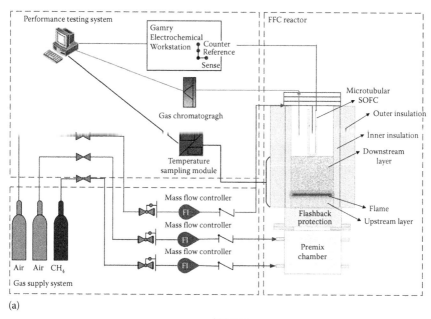

(a)

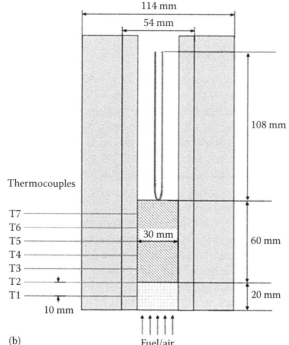

(b)

FIGURE 4.5 Schematic diagram of the FFC reactor system with the porous media burner. (a) Overview of the experimental system; (b) detailed configuration of the FFC reactor used by Wang et al. [42]. (Reprinted from *Energy*, 109, Wang, Y., Zeng, H., Shi, Y., Cao, T., Cai, N. et al., Power and heat co-generation by micro-tubular flame fuel cell on a porous media burner, 117–123, Copyright 2016, with permission from Elsevier.)

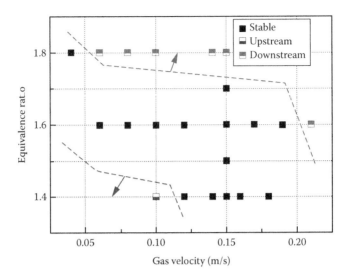

FIGURE 4.6 Flame stability limit for the porous media burner used by Wang et al. [42]. (Reprinted from *Energy*, 109, Wang, Y., Zeng, H., Shi, Y., Cao, T., Cai, N. et al., Power and heat co-generation by micro-tubular flame fuel cell on a porous media burner, 117–123, Copyright 2016, with permission from Elsevier.)

4.4.1 ELECTROCHEMICAL PERFORMANCE

Operational parameters such as equivalence ratios, inlet gas flow rates, and distances between the burner and the SOFC dominates the FFC performances. Temperature, composition, heat and mass transfer are also important factors of the electrochemical performances of FFCs. Overall, the temperature of SOFCs is extremely important to system performance and is affected by factors such as chemical combustion, heat conduction, and radiation heat loss [24]. Zhu [39] indicated that maximum power outputs are mainly affected by the temperature of SOFCs, whereas open-circuit voltages (OCVs) are not only affected by the temperature of SOFCs, but also by the gas composition of the anode and cathode. By increasing equivalence ratios past 1, concentrations of CO and H_2 in the exhaust gases of burners increase, resulting in FFC performance improvements. However, when equivalence ratios are too high, SOFC temperatures decrease significantly, leading to poor performance. When equivalence ratios are less than 1, production of CO and H_2 along with power output ceases.

Kronemayer et al. [24] studied the influences of equivalence ratios, gas velocities, and burner–SOFC distances on FFCs using fuels of methane, propane, and butane. By using a flat-flame burner and a panel SOFC, their experimental results showed that the power densities of FFCs increase with increasing equivalence ratios and gas velocities, along with decreasing burner–SOFC distances. Their results also showed, however, that maximum power density outputs did not occur at maximum equivalence ratios and gas velocities or minimum burner–SOFC distances. This is possibly attributed to the rich flame being unstable or unignitable at these conditions.

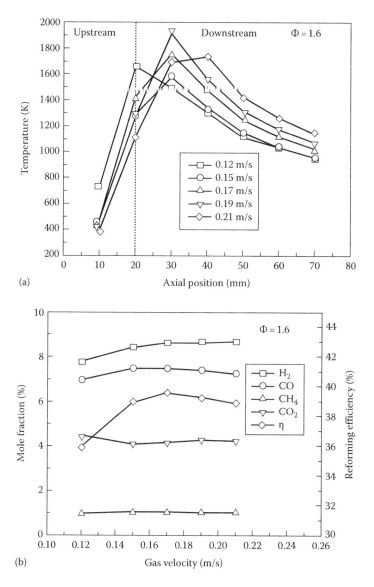

FIGURE 4.7 Flame temperature profiles (a) and compositions (b) at various velocities [42]. (Reprinted from *Energy*, 109, Wang, Y., Zeng, H., Shi, Y., Cao, T., Cai, N. et al., Power and heat co-generation by micro-tubular flame fuel cell on a porous media burner, 117–123, Copyright 2016, with permission from Elsevier.)

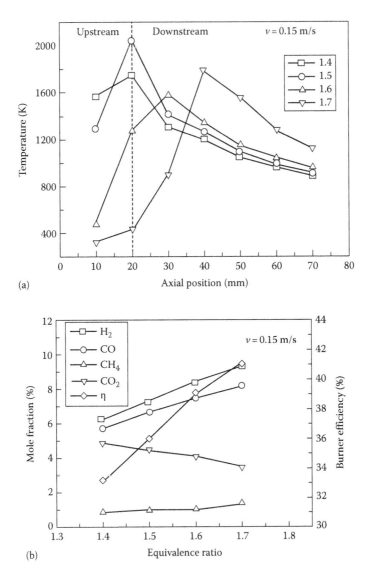

FIGURE 4.8 Flame temperature profiles (a) and compositions (b) at various equivalence ratios [42]. (Reprinted from *Energy*, 109, Wang, Y., Zeng, H., Shi, Y., Cao, T., Cai, N. et al., Power and heat co-generation by micro-tubular flame fuel cell on a porous media burner, 117–123, Copyright 2016, with permission from Elsevier.)

4.4.1.1 Effects of Equivalence Ratio

The equivalence ratio ϕ is defined as

$$\phi = \frac{n_{fuel}/n_{oxidant}}{n_{fuel}^{stoich}/n_{oxidant}^{stoich}} \quad (4.13)$$

where
   n represents the molar flow rates in the experiment
   n^{stoich} is the molar flow rates in stoichiometric conditions

The equivalence ratio greatly impacts FFC performances, mainly via the outlet gas composition from the burner.

Horiuchi et al. [21] demonstrated that fuel-to-air ratios can noticeably influence FFC performances. Vogler et al. [25] replaced the original Bunsen burner with a McKenna flat-flame burner and studied the performances of FFCs using equivalence ratios of 1.1 and 1.3 using methane as a fuel. The peak power density of their FFC reached 120 mW/cm^2 at an equivalence ratio of 1.3, with a generating efficiency of 0.45%.

Kronemayer et al. [24] explored the influence of equivalence ratios of methane/air flame on FFC performance, the investigation was carried out with a flat-flame burner and a 13 mm diameter panel SOFC. With an inlet gas velocity of 20 cm/s and a distance of 5–20 mm from the burner, the polarization curves of their SOFC were obtained under different equivalence ratios ranging from 1.1 to 1.4 (Figure 4.9). The relationships between maximum power density and equivalence ratios, inlet gas velocities, and burner–SOFC distances are summarized in Figure 4.9.

Wang et al. [41] integrated the microtubular SOFC with a two-layered porous media burner to study the electrochemical performances under various equivalence ratios. Their porous media burner consists of a 20 mm long upstream layer made up with 2–3 mm Al$_2$O$_3$ beads and a 60 mm long downstream layer consisting of 7.5 mm Al$_2$O$_3$ beads. The schematic of the experimental setup of Wang et al. is demonstrated in Figure 4.10. Their experimental results showed improvements in FFC performances (as shown in Figure 4.11a) along with decreases in electrochemical impedance when equivalence ratios are increased (as shown in Figure 4.11b). When the equivalence ratio is 1.6 and the inlet gas velocity is 0.15 m/s, the single cell reached 1.5 W at 0.7 V for 100 min.

4.4.1.2 Effects of Gas Flow Rate

Zhu et al. [20] studied the OCV of a microstack using a household cooker and liquefied petroleum gases. Figure 4.12 shows the OCV observed under various airflow rates at the cathode. When the inlet airflow rate is 150 sccm, an OCV of 0.9 V is obtained. If the inlet airflow rate is too low, the fuel at the anode side could easily diffuse to the cathode side, reducing the oxygen content at the cathode side, resulting in an OCV drop.

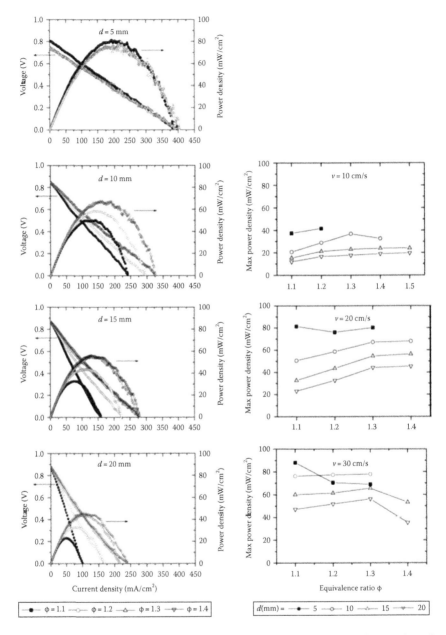

FIGURE 4.9 Left column: polarization curves of a DFFC system operated on methane for 20 cm s^{-1} gas in flow velocity and 5–20 mm distance from the burner for equivalence ratios φ = 1.1–1.4; Right column: Methane-operated DFFC: Maximum power density vs. equivalence ratio φ, distance between burner and SOFC d, and gas inlet velocity V [24]. (Reprinted from *J. Power Sources*, 166, Kronemayer, H., Barzan, D., Horiuchi, M., Suganuma, S., Tokutake, Y. et al., A direct-flame solid oxide fuel cell (DFFC) operated on methane, propane, and butane, 120–126, Copyright 2007, with permission from Elsevier.)

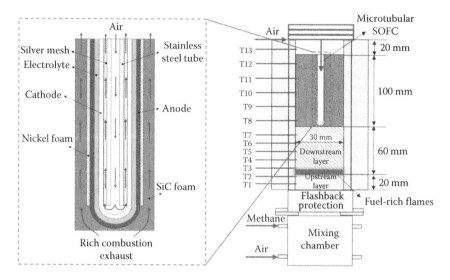

FIGURE 4.10 Schematic diagram for the FFC reactor used by Wang et al. [41]. (Reprinted from *Appl. Energy*, 178, Wang, Y., Zeng, H., Cao, T., Shi, Y., Cai, N. et al., Start-up and operation characteristics of a flame fuel cell unit, 415–421, Copyright 2016, with permission from Elsevier.)

Wang at al. [30] at Syracuse University used a quartz tube burner to study the influences of propane flow rate on FFC performance with a fixed airflow rate. The polarization curves and power densities are presented in Figure 4.13. Figure 4.14 illustrates the power densities with an airflow rate ranging from 0 to 500 mL/min and a fixed propane flow rate of 40 mL/min. With the fixed propane flow rate, the effects of airflow rates on polarization and power density are shown in Figure 4.15. These results demonstrate that the influences of airflow rates are complex when the temperature and concentration of reactants change. The total flow rate can significantly impact the temperature of SOFCs as well as the composition at the anode, resulting in changing maximum power densities.

Vogler et al. [25] published the polarization curves of methane/air FFCs at different inflow velocities with fixed fuel/air ratios of 1.1 and 1.3, using a flat-flame burner and a 13 mm diameter planar SOFC. Here, as the inflow velocity increases at a given equivalence ratio, the power density of the FFC increased significantly.

4.4.1.3 Effects of Burner–SOFC Distance

Wang et al. [37] in Nanjing University of Technology indicated that the relative locations between the SOFC and burner have great impacts on the temperature and the electrical performance of FFCs. Their experiments used an alcohol burner and demonstrated a maximum power density of 200 mW/cm² when the SOFC anode was mounted at the inner flame. The central flame of ethanol can be regarded as a fuel-rich flame having the highest concentration of hydrogen. It was observed that when heat released is insufficient; it would lead to lower

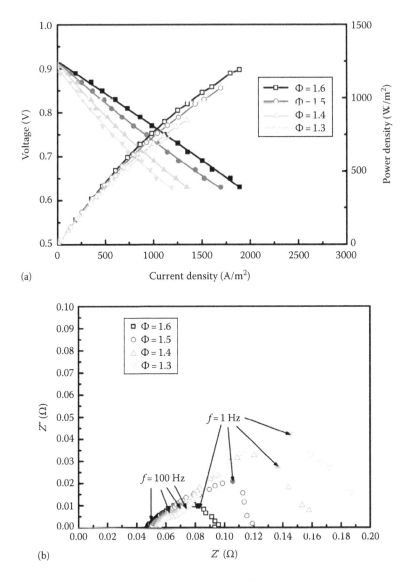

FIGURE 4.11 Electrochemical performance of the FFC with a porous combustor for various equivalence ratios. (a) Polarization curves. (b) EIS curves [41]. (Reprinted from *Appl. Energy*, 178, Wang, Y., Zeng, H., Cao, T., Shi, Y., Cai, N. et al., Start-up and operation characteristics of a flame fuel cell unit, 415–421, Copyright 2016, with permission from Elsevier.)

SOFC temperatures, decreased electrode reactivities and enlarged ohmic resistances. Although the temperature of the outer flame is highest, the concentration of hydrogen is insufficient. Comparing the central flame with the outer one, the temperature and composition of the central flame are more appropriate for operation of FFCs. Figure 4.16 shows the polarization and power density curves of different burner–SOFC distances.

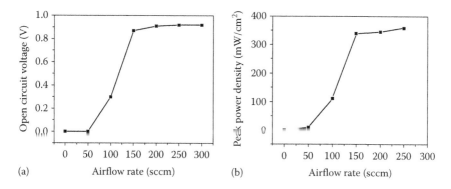

FIGURE 4.12 OCV (a) and peak power density (b) of a four-cell stack as a function of airflow rate on the cathode side [20]. (Reprinted from *Int. J. Hydrogen Energy*, 37, Zhu, X., Wei, B., Lü, Z., Yang, L., Huang, X. et al., A direct flame solid oxide fuel cell for potential combined heat and power generation, 8621–8629, Copyright 2012, with permission from Elsevier.)

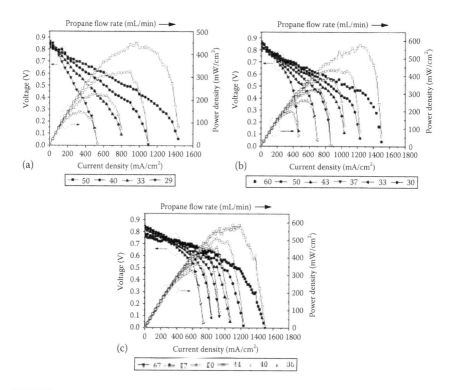

FIGURE 4.13 Polarization and power density curves of FFC with variations of propane flow rates at the airflow rate of (a) 200 mL/min; (b) 300 mL/min; and (c) 400 mL/min [30]. (Reprinted from *Proc. Combust. Inst.*, 33, Wang, K., Zeng, P., and Ahn, J., High performance direct flame fuel cell using a propane flame, 3431–3437, Copyright 2011, with permission from Elsevier.)

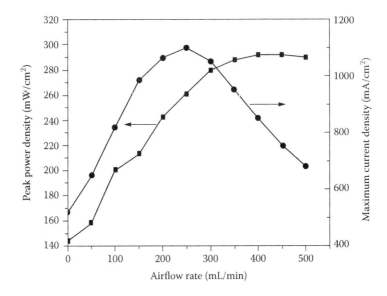

FIGURE 4.14 Peak power density and the maximum current density as a function of airflow rate at a propane flow rate of 40 mL/min [30]. (Reprinted from *Proc. Combust. Inst.*, 33, Wang, K., Zeng, P., and Ahn, J., High performance direct flame fuel cell using a propane flame, 3431–3437, Copyright 2011, with permission from Elsevier.)

Zhu et al. [20,39] indicated that open circuit voltages of FFCs can be affected by the distance between the FFCs and the burner. They found that the temperatures of the FFCs would be too low for operation when they are too far from the burner, but the oxygen partial pressures at the cathodes would be inadequate when they get too close to each other. Therefore, an optimized distance should be evaluated for stable operations of FFCs.

Sun et al. [17] in Nanjing University of Technology presented their polarization and power density curves and found that distances between the burner and FFCs could also influence the performances of the fuel cells as shown in Figure 4.17.

Zhu et al. [39] suggested that the effects of SOFC–burner distance on FFC performance is a consequence of the temperature variations on the FFCs. In their experiment, when the SOFC–burner distance was increased from 2 to 10 cm, the maximum power density of the FFC decreased. With an SOFC–burner distance of 2 cm, the temperature at the cathode was 880°C and the maximum power density was 238 mW/cm². The temperature of the SOFC would determine the conductivity of the electrolyte and the kinetics of the two electrodes.

4.4.2 FFC UNIT CONFIGURATIONS

The design of FFC unit is a key factor of fuel utilization efficiencies and operations environments of SOFCs, which would further affects the overall electrical efficiencies.

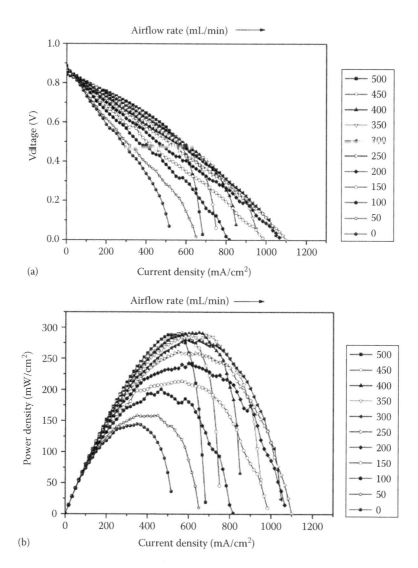

FIGURE 4.15 Polarization (a) and power density (b) curves at a propane flow rate of 40 mL/min [30]. (Reprinted from *Proc. Combust. Inst.*, 33, Wang, K., Zeng, P., and Ahn, J., High performance direct flame fuel cell using a propane flame, 3431–3437, Copyright 2011, with permission from Elsevier.)

Currently, several different FFC unit configurations have been proposed: the horizontal plate configuration, the vertical plate configuration, the horizontal tubular configuration, and the vertical tubular configuration, all four configurations are presented in Figure 4.18. The horizontal plate configuration is a simple structure and could be the most frequently adopted configuration in FFC studies. It is widely employed in FFC mechanism studies. Nonuniformity of fuel-rich flame on the surface of the FFC may cause local transient thermal stresses, leading to an SOFC failure. On the other

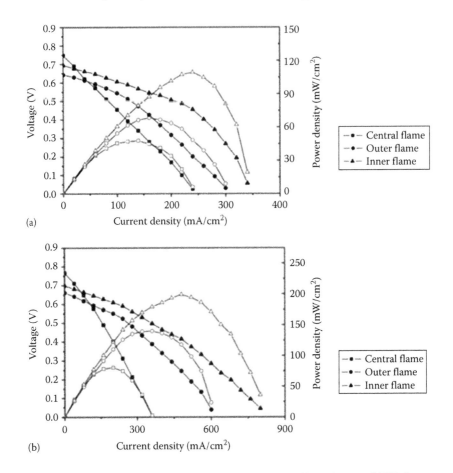

FIGURE 4.16 Polarization and power density curves with different burner–SOFC distances, before (a) and after (b) adding catalyst [37]. (Reprinted from *J. Power Sources*, 177, Wang, K., Ran, R., Hao, Y., Shao, Z., Jin, W., and Xu, N., A high-performance no-chamber fuel cell operated on ethanol flame, 33–39, Copyright 2008, with permission from Elsevier.)

hand, residence time of reactive gas in this configuration is too low for an acceptable fuel efficiency.

Horiuchi et al. [27] proposed the vertical plate configuration (Figure 4.18) in which a couple of fuel cells' anodes are put opposite to each other with cathodes being exposed to ambient air. The rich flame and reforming syngases are put into the channel formed by the anodes, extending the fuel gas residence time and increasing the power density of FFCs. It should be noted, however, that the limitations of quenching distance are serious barriers for higher power densities. For all hydrocarbon/air flames, the quenching distance is larger than 1 mm, and increases with increasing equivalence ratios. To minimize the quenching distance, Horiuchi et al. [27] replaced the open flame with a catalyst involved process in their actual experiments. In addition, due to the large temperature gradients at different heights, thermal management can be a great concern.

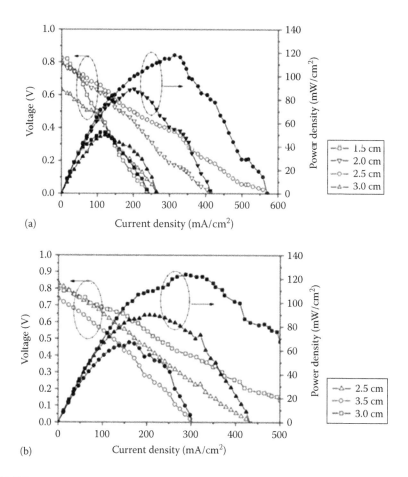

FIGURE 4.17 Polarization and power density curves affected by SOFC–burner distances with methanol (a) and ethanol (b) [17]. (Reprinted from *Int. J. Hydrogen Energy*, 35, Sun, L., Hao, Y., Zhang, C., Ran, R., and Shao, Z., Coking-free direct-methanol-flame fuel cell with traditional nickel–cermet anode, 7971–7981, Copyright 2010, with permission from Elsevier.)

The horizontal tubular configuration (Figure 4.18) was used by Wang et al. [43] to take the advantage of the thermal shock resistance of tubular SOFCs. Experimental observations proved that the nonuniformity in this configuration along the circumferential direction was small enough to maintain the FFC operation, leading to great advantage of thermal shock resistance over the horizontal plate configuration. Quick start up and easy thermal management are the major advantages for this kind of configuration. Relatively short reaction residence time of reactive gases is an important issue requires serious consideration.

The vertical tubular configuration (Figure 4.18) employed by Wang et al. [41,42] is also a reasonable configuration considering its flame uniformity, good thermal shock resistance, quick start-up, and long residence times. Since a single cell only

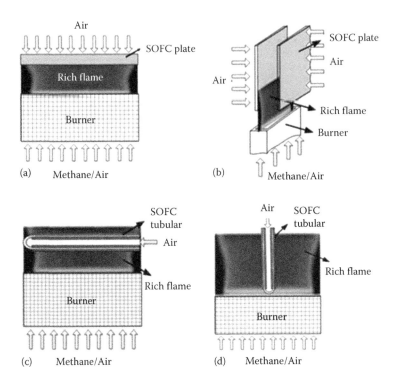

FIGURE 4.18 Common FFC unit configurations: (a) horizontal plate configuration [24]; (b) vertical plate configuration [27]; (c) horizontal tubular configuration; and (d) vertical tubular configuration [41,42].

occupies a small space, many tubular SOFCs can be connected in series or parallel upon a single burner to improve total power outputs. Large-scale FFCs units can be built based on this unit configuration.

With the vertical tubular configuration (Figure 4.18), an FFC stack, as shown in Figure 4.19, was successfully assembled and operated by Wang et al. [47] using methane as the fuel. The original configuration was expanded to a 36-tube stack, and could be scaled up further. A maximum power of 3.6 W for four microtubular SOFCs with an electrical efficiency of 6% and a fuel utilization of 23% was obtained. For microtubular SOFCs in various locations, uniform temperature distribution and gas composition were achieved using a porous combustor and a vertical tubular configuration.

4.5   CHALLENGES IN FFC

4.5.1   Thermal Shock Resistance

The operation of the combustor could induce thermal stresses within the electrode and electrolyte of the FFC, leading to serious mechanical problems [24]. Considering the operation conditions of FFCs, the heat-up rates of SOFCs are much faster compared

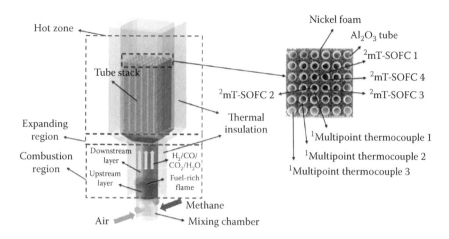

FIGURE 4.19 Schematic diagram of the FFC stack by Wang et al. [47]. (Reprinted from *Int. J. Hydrogen Energy*, Wang, Y., Shi, Y., Cao, T., Zeng, H., Cai, N. et al., A flame fuel cell stack powered by a porous media combustor, http://dx.doi.org/10.1016/j.ijhydene.2017.01.088, Copyright 2017, with permission from Elsevier.)

to conventional SOFC configurations. The FFCs could reach working temperatures within seconds [24]. In this case, the uniformity of the temperature field provided by the burner could not be guaranteed under certain circumstances, leading to severe thermal stress inside the SOFCs. These two aspects are the main differences between FFCs and conventional SOFCs.

Honda et al. [76] studied the creep rupture strength at elevated temperatures and the thermal shock resistance of Scandia-stabilized zirconia through piston on ring and infrared radiation heating. Their experimental results show that creep rupture strengths decline rapidly when temperatures are higher than 873 K, with its breaking strength at 1073 K being only half of that at room temperature. Thermal shock resistance is concerned with both the breaking strength and thermal expansion coefficient. In their experiment, the thermal shock resistance of ScSZ was 500–1500 W/m.

Wang et al. [40] built a two-dimensional direct FFC model to study heterogeneous and electrochemical reactions in the FFC unit. The model also evaluates the effects of electrode microstructures, transfer processes (charge, mass and heat), and thermal stresses on the fuel cell performance. The model was calculated via the finite element commercial software COMSOL Multiphysics®. This model is a useful tool to analyze temperature distributions and thermal stresses of SOFCs and is of great value for designing and optimizing FFCs. With the detailed two-dimensional model, the failure probability of FFCs and common SOFCs was compared. In addition, the thermal shock resistances of anode-supported SOFCs and electrode-supported SOFCs were also studied. The computational results showed that the thermal shock resistances of anode-supported SOFCs were smaller, and the failure possibility of electrode-supported SOFCs was two orders of magnitude larger than that of the anode-supported one. The results also suggested that the failure possibility of FFCs was six orders larger than that of traditional SOFCs, indicating that the uniformity of the fuel-rich

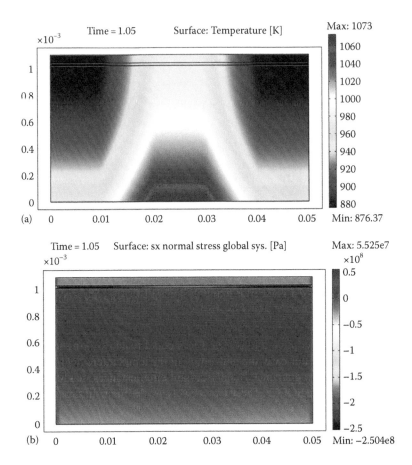

FIGURE 4.20 Transient temperature field (a) and transient stress field (b) in the start-up period [40]. (Reprinted from *J. Power Sources*, 255, Wang, Y., Shi, Y., Yu, X., and Cai, N., Thermal shock resistance and failure probability analysis on solid oxide electrolyte direct flame fuel cells, 377–386, Copyright 2014, with permission from Elsevier.)

flame is crucial to FFC operations. During the start-up period of FFCs, due to the rapid shifting temperature and flow field of the burner, temperature and stress distribution in the SOFC showed significant non-uniformity, and the calculated results are presented in Figure 4.20.

Considering the importance of thermal shock resistances in FFCs, Wang et al. [43] indicated that planar SOFC cell plates cracked easily when directly exposed to the flame environments during the start-up period. Figure 4.21 shows a picture of a fractured planar SOFC caused by the flame in the start-up period. Microtubular SOFCs proved themselves much more resistive to thermal shocks than the planar SOFCs [77] during the short start-up time (10 s ~ 10 min) [78,79]. Experiments from Wang et al. [43] demonstrated that microtubular SOFCs possess great start-up feasibilities and stabilities when integrated with a multielement diffusion flame burner (Hencken burner).

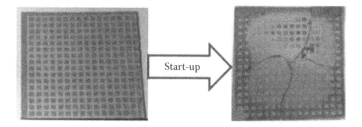

FIGURE 4.21   A planar FFC fractured in the start-up period.

4.5.2   CARBON DEPOSITION

Due to the large amounts of CO at the anode of FFCs, when the temperature is lower than the threshold value, the following exothermic reaction would take place:

$$2CO \rightleftharpoons CO_2 + C \quad \Delta H^0_{298} = -172.4 \text{ kJ/mol} \qquad (4.14)$$

This carbon deposition phenomenon may result in the significant degradation of FFC performances. After several cycles of discharging, anode concentration polarization losses also significantly increase due to the blocking effect of carbon deposition, further decreasing the power output of FFCs. Wang et al. [45] characterized the cross sectional morphologies of the anode with scanning electron microscopy (SEM) and the microscopic photo of the external surface of the anode is shown in Figure 4.22.

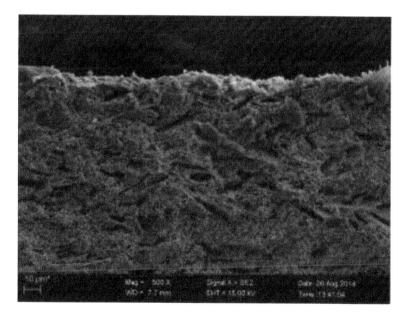

FIGURE 4.22   SEM morphology of the anode cross section near the surface. (Reproduced from Yuqing, W. et al., *J. Electrochem. Soc.*, 161, F1348, Copyright 2014, The Electrochemical Society.)

The observations under SEM and energy dispersive X-ray spectroscopy (EDX) show that carbon deposition mainly takes place at the external surface of the anode, blocking mass transport through the porous anode. The blockage effect of deposited carbon would increase the anode concentration polarization and decrease the FFC performance. Solutions to reduce carbon deposition were proposed, such as changing fuel types or improving anode materials.

4.5.2.1 Effects of Fuel Type

Sun et al. [17] studied the influence of fuel types on carbon deposition using a traditional Ni-SDC anode. Characterization methods such as SEM, EIS, $I–V$ curves, mass spectrum, and numerical simulations were adopted to study the carbon deposition and its impacts on FFC performance. The irreversible carbon deposition was observed when ethanol is fed as the fuel, and the performance of the FFC greatly decreased after three hours of operation. When methanol was used as the fuel, however, there was nearly zero carbon deposition after 30 hours of operation, with stable FFC performances.

4.5.2.2 Effects of Materials

The selection of materials, especially anode materials, is another determining factor of carbon deposition. Proper selection of anode materials and catalysts would help improve the performance of the SOFC, and suppress carbon deposition.

Wang et al. [37] reported that carbon deposition can be found on Ni-SDC anode, because Ni is a perfect catalyst for hydrocarbon decomposition. Power density of the fuel cell can be promoted by applying a thin Ru/SDC catalytic layer to the anode. Carbon deposition was found suppressed by the Ru based catalyst. They suggest that Ru catalysts can reduce carbon deposition by promoting the reaction of $C + H_2O \rightarrow CO + H_2$, therefore increasing its resistance towards carbon deposition. Noteworthy, the combination of Ru and ceria is a frequently reported catalytic system to eliminate carbon deposition.

Zhu et al. [20] studied the effects of anode materials on carbon deposition in FFCs. FFC unit consists of four single cells and a household cooker burning liquefied petroleum gases, they observed carbon deposition, performance instability, and long start-up times when employing Ni-YSZ as anode. However, when a composite anode mixed with $La_{0.75}Sr_{0.25}Cr_{0.5}Mn_{0.5}O_{3-\delta}$ (LSCM) was used, zero carbon deposition, stable operation, and quick start-up were achieved. They [39] also reported that carbon deposition under nonpremixed flames was much more serious than that under premixed flames. They addressed that by replacing the Ni-YSZ anode with LSCM impregnated anode, carbon deposition in the FFCs can be mitigated or eliminated.

4.6 FFC APPLICATIONS IN CHP SYSTEMS

Although fuel flexibility, simple configuration, rapid start-up, and feasible sealing make FFCs promising technologies for residential power generation their electrical efficiencies are relatively low compared to conventional SOFCs. However, for FFCs, it is important to consider both the electrical deficiency and the high grade heat from the combustor. For residential applications, heat and electricity are both necessary. A CHP power system built based on FFCs would fulfill both demands.

Wang et al. [46] proposed and analyzed a tri-generation system consisting of a direct FFC, a boiler, and a chiller based on the commercially available simulation platform gPROMS. The effects of equivalence ratio and fuel utilization on system efficiencies and thermal-to-electric ratios were studied. Figure 4.23 shows that system efficiencies can be increased by increasing equivalence ratios or fuel utilizations but at the cost of a reduction in thermal-to-electric ratios. Total system efficiencies

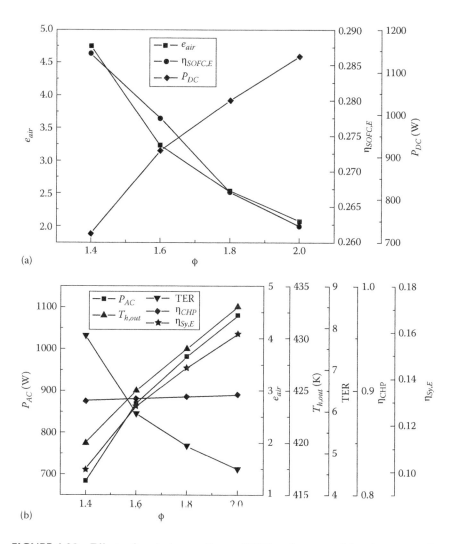

FIGURE 4.23 Effects of equivalence ratios on SOFC performances (a) and system performance (b) [46]. (Reprinted from *Int. J. Hydrogen Energy*, 39, Wang, Y., Shi, Y., Ni, M., Cai, N. et al., A micro tri-generation system based on direct flame fuel cells for residential applications, 5996–6005, Copyright 2014, with permission from Elsevier.)

TABLE 4.1
System Performance of Different Configurations [46]

System Configuration		Heat (W)	Cold (W)	Electricity (W)	System Efficiency (%)
Cogeneration	Heat + electricity	5124	0	922	89.2
	Cold + electricity	0	5615	922	96.5
Tri-generation	20% for heat + 80% for cold + electricity	1025	4329	922	93.4
	50% for heat + 50% for cold + electricity	2740	2562	922	91.9
	80% for heat + 20% for cold + electricity	4100	1096	922	90.3

Source: Reprinted from *Int. J. Hydrogen Energy*, 39, Wang, Y., Shi, Y., Ni, M., and Cai, N., A micro tri-generation system based on direct flame fuel cells for residential applications, 5996–6005. Copyright 2014, with permission from Elsevier.

can reach above 90% even though electrical efficiencies are only 20%, demonstrating the benefit of FFCs in a CHP system.

Based on energy demands of each family, Wang et al. [46] studied CHP generation in northern China along with combined cooling and power generation in southern China. They studied and analyzed various parameters of the tri-generation system, including SOFC fuel utilization factors, air excess ratios, SOFC stack electric efficiencies, system electric efficiencies, CHP efficiencies, CCHP efficiencies, and thermal-to-electric ratios. The results listed in Table 4.1 show that system efficiencies above 90% can be achieved in summer in Hong Kong and winter in Beijing.

Many researchers have proposed solutions yo promote the fuel utilization efficiencies for FFCs. Milcarek et al. [33] studied a micro-CHP system using a micro-tubular flame-assisted fuel cell to improve fuel utilization. Figure 4.24 shows a

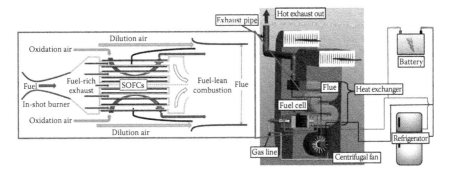

FIGURE 4.24 A residential-size flame-assisted fuel cell furnace for micro-CHP (FFF) concept [33]. (Reprinted from *J. Power Sources*, 306, Milcarek, R.J., Wang, K., Falkenstein-Smith, R.L., Ahn, J., Micro-tubular flame-assisted fuel cells for micro-combined heat and power systems, 148–151, Copyright 2016, with permission from Elsevier.)

flame-assisted fuel cell furnace using this micro-CHP (FFF) concept, which separates the SOFC from the combustion burner, allowing for increased electrochemical reaction areas. The mT-FFC under this configuration can achieve fuel utilizations of 29.1% and maximum power densities of 430 mW/cm².

4.7 SUMMARY

FFCs are novel energy systems combining SOFCs with fuel rich flames. The flame can provide both heat and fuel for the SOFC. The total electrical efficiency of the FFC system is determined by fuel-rich combustion efficiency, fuel utilization efficiency, and SOFC electrical efficiency. As one of the most important components in FFCs, the combustion burner has to produce uniform and stable fuel rich flames. The rich flammability limit has to be enlarged to produce adequate fuel (H_2 and CO) for the fuel cells. Equivalence ratios, inlet gas velocities, and burner–SOFC distances are the main factors affecting FFC performances, whereas thermal shock and carbon deposition are the main challenges. Combined heat and power generation systems with FFCs are promising for residential power supply.

REFERENCES

1. Pellegrino S., Lanzini A., Leone P. 2015. Techno-economic and policy requirements for the market-entry of the fuel cell micro-CHP system in the residential sector. *Applied Energy* 143: 370–382.
2. Hawkes A., Staffell I., Brett D., Brandon N. 2009. Fuel cells for micro-combined heat and power generation. *Energy & Environmental Science* 2: 729.
3. Loukou A., Frenzel I., Klein J., Trimis D. 2012. Experimental study of hydrogen production and soot particulate matter emissions from methane rich-combustion in inert porous media. *International Journal of Hydrogen Energy* 37: 16686–16696.
4. Brown J.E., Hendry C.N., Harborne P. 2007. An emerging market in fuel cells? Residential combined heat and power in four countries. *Energy Policy* 35: 2173–2186.
5. Walther D.C., Ahn J. 2011. Advances and challenges in the development of power-generation systems at small scales. *Progress in Energy and Combustion Science* 37: 583–610.
6. Singhal S.C. 2000. Advances in solid oxide fuel cell technology. *Solid State Ionics* 135: 305–313.
7. Obara S. 2010. Power generation efficiency of an SOFC–PEFC combined system with time shift utilization of SOFC exhaust heat. *International Journal of Hydrogen Energy* 35: 757–767.
8. Raimondi A., Loukou A., Fino D., Trimis D. 2011. Experimental analysis of soot abatement in reducing syngas for high temperature fuel cell feeding. *Chemical Engineering Journal* 176–177: 295–301.
9. Braun R.J., Klein S.A., Reindl D.T. 2006. Evaluation of system configurations for solid oxide fuel cell-based micro-combined heat and power generators in residential applications. *Journal of Power Sources* 158: 1290–1305.
10. Hawkes A., Leach M. 2005. Solid oxide fuel cell systems for residential micro-combined heat and power in the UK: Key economic drivers. *Journal of Power Sources* 149: 72–83.
11. Ormerod R.M. 2003. Solid oxide fuel cells. *Chemical Society Reviews* 32: 17–28.

12. Ellamla H.R., Staffell I., Bujlo P., Pollet B.G., Pasupathi S. 2015. Current status of fuel cell based combined heat and power systems for residential sector. *Journal of Power Sources* 293: 312–328.
13. Hibino T., Hashimoto A., Inoue T., Tokuno J., Yoshida S., Sano M. 2000. A low-operating-temperature solid oxide fuel cell in hydrocarbon-air mixtures. *Science* 288: 2031–2033.
14. Shao Z. 2004. Anode-supported thin-film fuel cells operated in a single chamber configuration 2T-I-12. *Solid State Ionics* 175: 39–46.
15. Suzuki T., Jasinski P., Petrovsky V., Anderson H.U., Dogan F. 2004. Anode supported single chamber solid oxide fuel cell in CH[sub 4]-air mixture. *Journal of the Electrochemical Society* 151: A1473.
16. Napporn T.W., Jacques-Bédard X., Morin F., Meunier M. 2004. Operating conditions of a single-chamber SOFC. *Journal of the Electrochemical Society* 151: A2088.
17. Sun L., Hao Y., Zhang C., Ran R., Shao Z. 2010. Coking-free direct-methanol-flame fuel cell with traditional nickel–cermet anode. *International Journal of Hydrogen Energy* 35: 7971–7981.
18. Yano M., Tomita A., Sano M., Hibino T. 2007. Recent advances in single-chamber solid oxide fuel cells: A review. *Solid State Ionics* 177: 3351–3359.
19. Shao Z.P., Haile S.M. 2004. A high-performance cathode for the next generation of solid-oxide fuel cells. *Nature* 431: 170–173.
20. Zhu X. et al. 2012. A direct flame solid oxide fuel cell for potential combined heat and power generation. *International Journal of Hydrogen Energy* 37: 8621–8629.
21. Horiuchi M., Suganuma S., Watanabe M. 2004. Electrochemical power generation directly from combustion flame of gases, liquids, and solids. *Journal of the Electrochemical Society* 151: A1402.
22. Nakamura Y., Endo S. 2015. Power generation performance of direct flame fuel cell (DFFC) impinged by small jet flames. *Journal of Micromechanics and Microengineering* 25: 104015.
23. Endo S., Nakamura Y. 2014. Power generation properties of Direct Flame Fuel Cell (DFFC). *Journal of Physics Conference Series* 557: 012119.
24. Kronemayer H. et al. 2007. A direct-flame solid oxide fuel cell (DFFC) operated on methane, propane, and butane. *Journal of Power Sources* 166: 120–126.
25. Vogler M. et al. 2007. Direct-flame solid-oxide fuel cell (DFFC): A thermally self-sustained, air self-breathing, hydrocarbon-operated SOFC system in a simple, no-chamber setup. *ECS Transactions* 7: 555–564.
26. Vogler M., Horiuchi M., Bessler W.G. 2010. Modeling, simulation and optimization of a no-chamber solid oxide fuel cell operated with a flat-flame burner. *Journal of Power Sources* 195: 7067–7077.
27. Horiuchi M. et al. 2009. Performance of a solid oxide fuel cell couple operated via in situ catalytic partial oxidation of n-butane. *Journal of Power Sources* 189: 950–957.
28. Wang K. 2015. Flame-assisted fuel cell operating with methane for combined heating and micro power. In *Proceedings of the ASME 2015 13th International Conference on Fuel Cell Science, Engineering and Technology FUEL CELL 2015*, June 28–July 2, 2015, San Diego, CA.
29. Wang K., Milcarek R.J., Zeng P., Ahn J. 2015. Flame-assisted fuel cells running methane. *International Journal of Hydrogen Energy* 40: 4659–4665.
30. Wang K., Zeng P., Ahn J. 2011. High performance direct flame fuel cell using a propane flame. *Proceedings of the Combustion Institute* 33: 3431–3437.
31. Falkenstein-Smith R., Wang K., Milcarek R., Ahn J. 2016. Integrated anaerobic digester and fuel cell power generation system for community use. *ASME 2015 13th International Conference on Fuel Cell Science, Engineering and Technology Collocated*

with the *ASME 2015 Power Conference, the ASME 2015 9th International Conference on Energy Sustainability, and the ASME 2015 Nuclear Forum.* American Society of Mechanical Engineers: V001T03A004, June 28–July 2, 2015, San Diego, CA.

32. Milcarek R.J., Garrett M.J., Ahn J. 2016. Micro-tubular flame-assisted fuel cell stacks. *International Journal of Hydrogen Energy* 41: 21489–21496.

33. Milcarek R.J., Wang K., Falkenstein-Smith R.L., Ahn J. 2016. Micro-tubular flame-assisted fuel cells for micro-combined heat and power systems. *Journal of Power Sources* 306: 148–151.

34. Milcarek R.J., Wang K., Garrett M.J., Ahn J. 2016. Micro-tubular flame-assisted fuel cells running methane. *International Journal of Hydrogen Energy* 41: 20670–20679.

35. Milcarek R.J., Garrett M.J., Wang K., et al. 2015. Performance investigation of dual layer yttria-stabilized zirconia–samaria-doped ceria electrolyte for intermediate temperature solid oxide fuel cells. *Journal of Electrochemical Energy Conversion and Storage* 13: 011002.

36. Aguilar-Arias J., Hotza D., Lenormand P., Ansart F. 2013. Planar solid oxide fuel cells using PSZ, processed by sequential aqueous tape casting and constrained sintering. *Journal of the American Ceramic Society* 96: 3075–3083.

37. Wang K., Ran R., Hao Y., Shao Z., Jin W., Xu N. 2008. A high-performance no-chamber fuel cell operated on ethanol flame. *Journal of Power Sources* 177: 33–39.

38. Wang Y., Sun L., Luo L., Wu Y., Liu L., Shi J. 2014. The study of portable direct-flame solid oxide fuel cell (DF-SOFC) stack with butane fuel. *Journal of Fuel Chemistry and Technology* 42: 1135–1139.

39. Zhu X., Lü Z., Wei B. et al. 2010. Direct flame SOFCs with $La_{0.75}Sr_{0.25}Cr_{0.5}Mn_{0.5}O_{3-\delta}$/Ni coimpregnated yttria-stabilized zirconia anodes operated on liquefied petroleum gas flame. *Journal of the Electrochemical Society* 157: B1838–B1843.

40. Wang Y., Shi Y., Yu X., Cai N. 2014. Thermal shock resistance and failure probability analysis on solid oxide electrolyte direct flame fuel cells. *Journal of Power Sources* 255: 377–386.

41. Wang Y. et al. 2016. Start-up and operation characteristics of a flame fuel cell unit. *Applied Energy* 178: 415–421.

42. Wang Y. et al. 2016. Power and heat co-generation by micro-tubular flame fuel cell on a porous media burner. *Energy* 109: 117–123.

43. Wang Y., Shi Y., Cai N., Ye X., Wang S. 2015. Performance characteristics of a micro-tubular solid oxide fuel cell operated with a fuel-rich methane flame. *ECS Transactions* 68: 2237–2243.

44. Wang Y.Q., Shi Y.X., Yu X.K., Cai N.S., Li S.Q. 2013. Integration of solid oxide fuel cells with multi-element diffusion flame burners. *Journal of the Electrochemical Society* 160: F1241–F1244.

45. Yuqing W., Yixiang S., Xiankai Y., Ningsheng C., Jiqing Q., Shaorong W. 2014. Experimental characterization of a direct methane flame solid oxide fuel cell power generation unit. *Journal of the Electrochemical Society* 161: F1348–F1353.

46. Wang Y., Shi Y., Ni M., Cai N. 2014. A micro tri-generation system based on direct flame fuel cells for residential applications. *International Journal of Hydrogen Energy* 39: 5996–6005.

47. Wang Y. et al. 2017. A flame fuel cell stack powered by a porous media combustor. *International Journal of Hydrogen Energy.* http://dx.doi.org/10.1016/j.ijhydene.2017.01.088.

48. Hossain M.M. et al. 2015. Study on direct flame solid oxide fuel cell using flat burner and ethylene flame. *ECS Transactions* 68: 1989–1999.

49. Speelman N., Kiefer M., Markus D., Maas U., de Goey L.P.H., van Oijen J.A. 2015. Validation of a novel numerical model for the electric currents in burner-stabilized methane–air flames. *Proceedings of the Combustion Institute* 35: 847–854.

50. Senser D.W., Morse J.S., Cundy V.A. 1985. Construction and novel application of a flat flame burner facility to study hazardous waste combustion. *Review of Scientific Instruments* 56: 1279.

51. Cheskis S. 1999. Quantitative measurements of absolute concentrations of intermediate species in flames. *Progress in Energy and Combustion Science* 25: 233–252.

52. Chen Y.L., Lewis J., Parigger C. 2000. Probability distribution of laser-induced breakdown and ignition of ammonia. *Journal of Quantitative Spectroscopy and Radiative Transfer* 66: 41–53.

53. Meier W., Vyrodov A.O., Bergmann V., Stricker W. 1996. Simultaneous Raman/LIF measurements of major species and NO in turbulent H-2/air diffusion flames. *Applied Physics B-Lasers and Optics* 63: 79–90.

54. Migliorini F., Deiuliis S., Cignoli F., Zizak G. 2008. How "flat" is the rich premixed flame produced by your McKenna burner? *Combustion and Flame* 153: 384–393.

55. Ma J., Fletcher T.H., Webb, B.W. 1996. Conversion of coal tar to soot during coal pyrolysis in a post flame environment. In *Symposium (International) on Combustion*, vol. 26(2), July 28–August 2, 1996, Naples, Italy. pp. 3161–3167.

56. Ma J., Fletcher T.H., Webb B.W. 1995. Thermophoretic sampling of coal-derived soot particles during devolatilization. *Energy and Fuels* 9: 802–808.

57. Molina A., Shaddix C.R. 2007. Ignition and devolatilization of pulverized bituminous coal particles during oxygen/carbon dioxide coal combustion. *Proceedings of the Combustion Institute* 31: 1905–1912.

58. Murphy J.J., Shaddix C.R. 2006. Combustion kinetics of coal chars in oxygen-enriched environments. *Combustion and Flame* 144: 710–729.

59. Shaddix C.R., Molina A. 2009. Particle imaging of ignition and devolatilization of pulverized coal during oxy-fuel combustion. *Proceedings of the Combustion Institute* 32: 2091–2098.

60. van Eyk P.J., Ashman P.J., Alwahabi Z.T., Nathan G.J. 2008. Quantitative measurement of atomic sodium in the plume of a single burning coal particle. *Combustion and Flame* 155: 529–537.

61. van Eyk P.J., Ashman P.J., Alwahabi Z.T., Nathan G.J. 2009. Simultaneous measurements of the release of atomic sodium, particle diameter and particle temperature for a single burning coal particle. *Proceedings of the Combustion Institute* 32: 2099–2106.

62. Wooldridge M.S. et al. 2002. An experimental investigation of gas-phase combustion synthesis of SiO_2 nanoparticles using a multi-element diffusion flame burner. *Combustion and Flame* 131: 98–109.

63. Zhdanok S., Kennedy L.A., Koester G. 1995. Superadiabatic combustion of methane air mixtures under filtration in a packed bed. *Combustion and Flame* 100: 221–231.

64. Bingue J.P., Saveliev A.V., Fridman A.A., Kennedy L.A. 2002. Hydrogen production in ultra-rich filtration combustion of methane and hydrogen sulfide. *International Journal of Hydrogen Energy* 27: 643–649.

65. Mujeebu M.A., Abdullah M.Z., Abu Bakar M.Z., Mohamad A.A., Muhad R.M.N., Abdullah M.K. 2009. Combustion in porous media and its applications: A comprehensive survey. *Journal of Environmental Management* 90: 2287–2312.

66. Mujeebu M.A., Abdullah M.Z., Abu Bakar M.Z., Mohamad A.A., Abdullah M.K. 2009. Applications of porous media combustion technology—A review. *Applied Energy* 86: 1365–1375.

67. Wood S., Harris A.T. 2008. Porous burners for lean-burn applications. *Progress in Energy and Combustion Science* 34: 667–684.

68. Wang H., Wei C., Zhao P., Ye T. 2014. Experimental study on temperature variation in a porous inert media burner for premixed methane air combustion. *Energy* 72: 195–200.

69. Belmont E.L., Solomon S.M., Ellzey J.L. 2012. Syngas production from heptane in a non-catalytic counter-flow reactor. *Combustion and Flame* 159: 3624–3631.

70. Toledo M., Bubnovich V., Saveliev A., Kennedy L. 2009. Hydrogen production in ultrarich combustion of hydrocarbon fuels in porous media. *International Journal of Hydrogen Energy* 34: 1818–1827.
71. Bingue J.P., Saveliev A., Kennedy L.A. 2004. Optimization of hydrogen production by filtration combustion of methane by oxygen enrichment and depletion. *International Journal of Hydrogen Energy* 29: 1365–1370.
72. Drayton M.K., Saveliev A.V., Kennedy L.A., Fridman A.A., Li Y.E. 1998. Syngas production using superadiabatic combustion of ultra-rich methane-air mixtures. In *Twenty-Seventh Symposium (International) on Combustion*, vol. 27, August 2–7, 1998, Boulder, CO, pp. 1361–1367.
73. Dhamrat R.S., Ellzey J.L. 2006. Numerical and experimental study of the conversion of methane to hydrogen in a porous media reactor. *Combustion and Flame* 144: 698–709.
74. Pedersen-Mjaanes H., Chan L., Mastorakos E. 2005. Hydrogen production from rich combustion in porous media. *International Journal of Hydrogen Energy* 30: 579–592.
75. Al-Hamamre Z., Voss S., Trimis D. 2009. Hydrogen production by thermal partial oxidation of hydrocarbon fuels in porous media based reformer. *International Journal of Hydrogen Energy* 34: 827–832.
76. Honda S. et al. 2009. Strength and thermal shock properties of scandia-doped zirconia for thin electrolyte sheet of solid oxide fuel cell. *Materials Transactions* 50: 1742–1746.
77. Kendall K. 2010. Progress in microtubular solid oxide fuel cells. *International Journal of Applied Ceramic Technology* 7: 1–9.
78. Lockett M., Simmons M.J.H., Kendall K. 2004. CFD to predict temperature profile for scale up of micro-tubular SOFC stacks. *Journal of Power Sources* 131: 243–246.
79. Howe K.S., Thompson G.J., Kendall K. 2011. Micro-tubular solid oxide fuel cells and stacks. *Journal of Power Sources* 196: 1677–1686.

5 Solid Oxide Direct Carbon Fuel Cell

5.1 INTRODUCTION

Direct carbon fuel cells (DCFCs) convert solid carbon fuels into power through electrochemical reactions at high temperatures. Instead of gaseous or liquid fuels required by other types of fuel cells, DCFCs can use a wide variety of solid carbon materials as fuels, including carbon, carbon black, biomass char, and municipal waste. Compared with traditional coal-fired power plants that employ the Rankine cycle to generate power, DCFCs possess unique advantages such as high efficiencies, abundant fuel supplements, low fuel-processing costs, and excellent pollution controls.

A DCFC is made up of two electrodes and an ionic-conducting electrolyte. At the anode, solid carbon fuels and other intermediate reducing agents are converted through a carbon fuel conversion process ($C + 2O^{2-} \rightarrow CO_2 + 4e^-$) and supply electrons for the overall electrochemical reaction. At the cathode, oxygen reduction reactions take place to accept electrons from the anode side to produce O^{2-} ($O_2 + 4e^- \rightarrow 2O^{2-}$). Between these two electrodes, there is a solid oxide electrolyte functioning to transport O^{2-} ions to complete the reaction cycle and to insulate the cathode chamber from the anode chamber, blocking electron transport between the two electrodes. The total reaction of the cell is $C + O_2 \rightarrow CO_2$. The electrode potential at the cathode is higher than that of the anode, and this potential difference defines the cell voltage and drives the electron flow from the anode to the cathode to produce current and electrical power. Despite CO_2 being produced, DCFCs are more environmentally friendly than traditional carbon-burning technologies. This is mainly due to its higher efficiencies, requiring less carbon to produce the same amount of energy. Furthermore, because emissions of DCFC reactions consist of only pure carbon dioxide, carbon capture techniques are much cheaper than those of conventional power stations.

For fuel cells in general, solid carbon fuels possess significantly higher power densities than gas or liquid fuels and are widely available in large quantities and low prices. There are obvious disadvantages to solid carbon fuels however, such as low reaction kinetics as well as difficulties in transportation to electrochemically reactive sites at the anode. Therefore, most DCFCs operate at elevated temperatures ($600°C–1000°C$) in order to activate carbon-converting processes. Efforts have also been made to covert solid carbon fuels into gaseous fuels as well as to transport oxygen-containing species to the carbon fuels instead. Carbon conversion mechanisms play an important role in increasing the carbon reaction rates of DCFCs.

One of the most fatal problems with high-temperature fuel cells fed with carbonaceous fuels is anode coking (carbon deposit formation), which blackens reaction active sites. Proper control of carbon deposition and conversion through the optimization of anode materials, microstructures and surface conditions are required to enhance fuel cell performances, increase efficiencies, and prolong lifetimes of anodes.

Although there are prospective practical applications of DCFC technologies in energy conversion, there are still unresolved application issues in high-temperature DCFCs. These issues include the effects of metals contained in carbon fuels on fuel cell performances, the management of ash left over by carbon conversion, the development of advanced carbon conversion electrodes, and the proper configuration of solid carbon fueled DCFC systems. To facilitate DCFC technology research and development, this chapter will review the latest progress in advanced electrode materials, electrode/cell designs, microscopic structures, the fundamental understandings of DCFC principles and reaction mechanisms of high-temperature electrochemical carbon conversions. Technical challenges will also be discussed, and possible research directions to overcome these challenges will be proposed.

5.2 THERMODYNAMICS OF CARBON CONVERSION

5.2.1 OPEN CIRCUIT POTENTIAL OF DCFCS

Open circuit voltage (OCV), an important parameter of DCFCs, describes the chemical potential difference between the anode and cathode. The anode and cathode and the overall DCFC reaction can be described by Equations 5.1 through 5.3:

$$C + 2O^{2-} \rightarrow CO_2 + 4e^- \quad \text{(Anode reaction)} \tag{5.1}$$

$$O_2 + 4e^- \rightarrow 2O^{2-} \quad \text{(Cathode reaction)} \tag{5.2}$$

$$O_2 + C \rightarrow CO_2 \quad \text{(Overall reaction)} \tag{5.3}$$

The thermodynamic cell voltage (or Nernst voltage, E) of DCFCs can be calculated from the oxygen partial pressure difference between the electrodes, as shown as Equation 5.4:

$$E = \frac{RT}{nF} \ln \left(\frac{P_{O_2,\text{ca}}}{P_{O_2,\text{an}}} \right) \tag{5.4}$$

where
E denotes the OCV of a DCFC
$P_{O_2,\text{ca}}$ and $P_{O_2,\text{an}}$ are the values of oxygen partial pressure in the cathode and anode
R is the ideal gas constant (8.314 J mol^{-1} K^{-1})
T is the operation temperature (K)
F is Faraday's constant (96485.3 C mol^{-1})
n is the molar number of electrons transferred per mole oxygen consumed (mol/mol)

Considering the overall reaction of complete carbon oxidation (Equation 5.3), where $n = 4$, and the oxygen partial pressure in the anode chamber is determined by the Van't Hoff equation:

$$\ln\left(\frac{P_{CO_2}}{P_{O_2,an}\gamma_C}\right) = -\frac{\Delta G_C}{RT} \tag{5.5}$$

where

P_{CO_2} denotes the CO_2 partial pressure in the anode chamber

ΔG_C denotes the Gibbs free energy change of reaction (5.3) (J/mol) at a given temperature

γ_C denotes the reactivity of carbon

Open circuit potential (OCP) can be expressed as

$$E = \frac{RT}{nF}\ln\left(P_{O_2,ca}\right) - \frac{\Delta G_C}{nF} - \frac{RT}{nF}\ln\left(\frac{P_{CO_2}}{\gamma_C}\right) \tag{5.6}$$

If air is employed as the oxidant, the oxygen partial pressure at the cathode is 0.21, and the values of P_{CO_2}, as in Equation 5.3 , are chosen to be unity.

In reality, the OCV value of DCFCs is also affected by other reactions at the anode:

$$C + O^{2-} \rightarrow CO + 2e^- \tag{5.7}$$

$$CO + O^{2-} \rightarrow CO_2 + 2e^- \tag{5.8}$$

The produced CO_2 from Equation 5.6 can chemically react with solid C through Equation 5.9:

$$C + CO_2 \rightarrow 2CO \tag{5.9}$$

At different temperatures, various concentrations of CO at the anode chamber can affect the OCV, as shown in Figure 5.1, in which the OCV values of complete carbon oxidation are expressed as the shadowed area between the black solid lines with the reactivity of various types of carbon fuels earlier. The other shadowed region located between the red dashed lines shows the OCV values of the partial oxidation of carbon, which is closely related to the reactivity of the solid carbon fuel. The OCV values of CO oxidation are shown as the blue solid line in the figure. Under many circumstances, CO is one of the most important intermediates of high-temperature electrochemical conversion of carbon.

For the reactivity of carbon (γ_C) in Equation 5.6, Li et al. [1] proposed a method to calculate the value as a ratio between the number of activated carbon atoms and the total number of carbon atoms based on a "perfect graphite" assumption, in which the carbon fuel is made up of layers of round, paralleled perfect graphene crystallites of

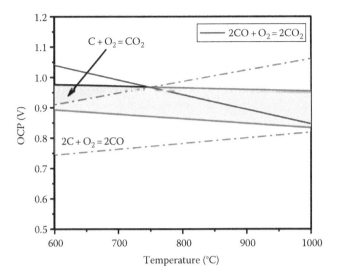

FIGURE 5.1 OCV values of carbon oxidation and CO-related reactions as a function of temperature.

the same diameter, noted as $L\alpha$. These two-dimensional graphene layers, as shown in Figure 5.2, are composed of hexagonal carbon rings with l as the distance between neighboring carbon atoms.

The number of carbon atoms (N_{tot}) in a single crystallite can be calculated by Equation 5.10:

$$N_{tot} = \frac{0.25 \times \pi \times L\alpha^2}{\frac{3\sqrt{3}}{2} \times l^2} \times 2 \tag{5.10}$$

Regarding the activated carbon atoms, only the ones located at the edges and defects of the crystallite are assumed to be reactive. Under the "perfect graphite" assumption, they are distributed evenly only on the edge of the round graphene crystallite. Based on this, the number of reactive carbon atoms (N_{act}) can be calculated through Equation 5.11:

$$N_{act} = \frac{\pi \times L\alpha}{\sqrt{3} \times l} \tag{5.11}$$

In Equation 5.11, the numerator term represents the perimeter of the graphene crystallite, or more briefly, the length of the edge, whereas the denominator denotes the distance between two neighboring active carbon atoms. By combining Equations 5.10 and 5.11, the activity of carbon can be acquired by Equation 5.12:

$$\gamma_C = \frac{N_{act}}{N_{tot}} = \frac{3l}{L\alpha} \tag{5.12}$$

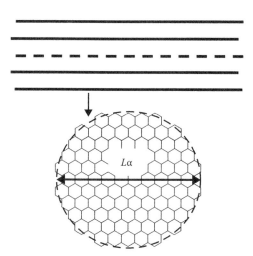

FIGURE 5.2 Sketch of layers of graphene crystallites. (Reprinted from *Electrochim. Acta*, 55, Li, H., Liu, Q., and Li, Y., A carbon in molten carbonate anode model for a direct carbon fuel cell, 1958–1965, Copyright 2010, with permission from Elsevier.)

For hexagonal carbon rings, $l = 0.142$ nm, while the value of $L\alpha$ varies for different types of carbon. For carbon black, $L\alpha$ ranges from 1.2 to 5 nm, whereas for graphite, $L\alpha$ ranges from 10 to 100 nm. This "perfect graphite" assumption demonstrates the influence of carbon microstructures on the OCV value of DCFCs. With different carbon materials, the influences of γ_C on OCV range from 0.02 to 0.12 V. Equations 5.11 and 5.12 indicate that a larger γ_C can lead to a higher OCV value, suggesting that carbons with smaller $L\alpha$ values and more defects are more favorable for DCFCs. This is also true for carbon conversion kinetics.

5.2.2 THEORETICAL EFFICIENCY OF DCFCs

As discussed in a previous chapter, the theoretical efficiency of a fuel cell is defined as the ratio between the Gibbs free energy change (ΔG) and the enthalpy change (ΔH) of fuel oxidation as shown in Equation 5.13, where ΔS is the entropy change of the given reaction:

$$\eta_{theo} = \frac{\Delta G}{\Delta H} \times 100\% = \left(1 - T\frac{\Delta S}{\Delta H}\right) \times 100\% \qquad (5.13)$$

Figure 5.3 shows that the theoretical efficiency of gaseous fuels is no more than 80% at high temperatures, while the efficiency value of solid carbon fuels, represented by the shadowed zone between the solid black lines, is much higher. This high electrochemical carbon conversion efficiency is a result of the small entropy change of the carbon oxidation reaction.

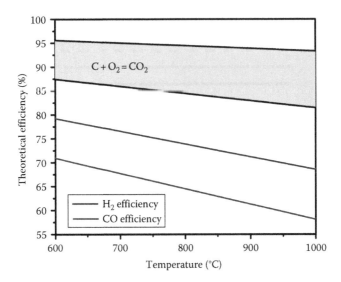

FIGURE 5.3 Theoretical efficiency of carbon oxidation in contrast with H_2 and CO as a function of temperature.

5.2.3 PRACTICAL EFFICIENCY OF DCFCS

The earlier discussions of theoretical efficiency are based on a thermodynamic point of view, in which the fuel cell system is assumed to have no net current passing through. In actual operations however, substantial currents will be drawn from the DCFC system, in which the cell voltage will become lower than that of the OCV due to several polarization processes occurring inside the cell. This cell voltage loss is mainly attributed to overpotentials induced by both anode and cathode reactions, reactant transport limitations, and electrolyte resistances. This voltage loss can significantly decrease the overall DCFC efficiency (η_{real}), as expressed by Equation 5.14:

$$\eta_{real} = \eta_{theo} \times \eta_{vol} \times \eta_{fuel} \tag{5.14}$$

where η_{vol} denotes the voltage efficiency and can be expressed as a ratio of the operation voltage (V) over the OCV value (E), shown by Equation 5.15:

$$\eta_{vol} = \frac{V}{E} \times 100\% \tag{5.15}$$

η_{fuel} is the fuel efficiency, which is the ratio between the fuel involved in the fuel cell reaction (ν_{invo}, mol/s) and the fuel introduced into the system (ν_{fed}, mol/s), shown by Equation 5.16:

$$\eta_{fuel} = \frac{\nu_{invo}}{\nu_{fed}} \times 100\% \tag{5.16}$$

For fuel cells using purely H_2 or CO as a fuel, there is only one type of oxidation product. In this case, the rate of fuel consumption can be identified as a function of the current drawn from the fuel cell: $v_{invo} = \dfrac{i}{nF}$, where i is the current measured during fuel cell operations, n is the mole value of the electrons transferred per mole of fuel oxidized, and F is Faraday's constant. In the case of carbon-oxidizing systems, oxidized products contain both CO and CO_2 however, and practical efficiencies need to be calculated based on flue gas compositions.

When both gas compositions and electrochemical data have been collected, practical efficiencies can be evaluated by Equation 5.17:

$$\eta_{real} = \frac{W_{el}}{Q_{fuel}} \times 100\% \tag{5.17}$$

where

W_{el} is the electrical work done by the fuel cell

Q_{fuel} is the chemical energy consumed during the fuel cell operation

W_{el} is calculated based on the electrochemical data acquired during the fuel cell test, an integration of work done over time, shown by $W_{el} = \int iUdt$; U is the working potential, and i is the current value. Both U and i values are monitored and recorded as a function of time during fuel cell operations. Q_{fuel} is determined based on heat values and the mass of carbon fuels introduced into the anode system, expressed by $Q_{fuel} = (LHV \text{ or } HHV)_{fuel,in} m_{fuel,in} - (LHV \text{ or } HHV)_{fuel,out} m_{fuel,out}$, where LHV and HHV are the heat values of the carbon fuel, and $m_{fuel,in}$ is the carbon loaded into the anode chamber of the DCFC. The second term on the right of the Q_{fuel} expression denotes the energy contained in the fuel escaping from the DCFC reactor. Because solid carbon fuels cannot be carried out of the anode chamber by generated flue gases, the energy escaping out of the anode chamber is actually the heat value of the CO-containing flue gases, which is a function of the gas composition. Because the flue gases of a DCFC system contain various reducing species produced by carbon fuel oxidation or pyrolysis, the definition of practical efficiency shown in Equation 5.17 is also applicable for fuel cell systems working under varying loads and fueled by mixed fuels.

The calculations presented earlier demonstrate that the thermodynamic parameters of a DCFC system are not affected by the electrode or electrolyte materials and are only determined by the temperature and the properties of the fuels and oxidants applied to the DCFC system.

5.3 DCFC CONFIGURATIONS

As mentioned previously, DCFCs often require elevated temperatures to operate. Therefore, high-temperature components are necessary for carbon conversion, including solid oxides, molten hydroxides, molten carbonates, and molten metals.

5.3.1 Solid Oxide DCFCs

Solid oxide fuel cells (SOFCs) are being intensively studied as a type of high-temperature fuel cell. SOFCs employ solid-state, ZrO_2-based (e.g., YSZ, yttria-stabilized zirconia) or CeO_2-based (e.g., GDC, gadolinia-doped ceria) ionic-conducting ceramics as electrolytes [2] and nickel-based porous cermets containing nickel and ionic-conducting ceramics as anodes, along with porous perovskite-type ceramics as cathodes. These SOFCs generally operate at temperatures between 600°C and 1000°C, and are capable of oxidizing various types of carbon containing fuels. Therefore, SOFCs can be regarded as solid-oxide-based DCFCs (SO-DCFC) when solid carbon fuels are directly introduced into their anode chambers. The working process and configuration of an SO-DCFC are shown in Figure 5.4, in which pulverized carbon is fed directly into the anode chamber, forming a carbon bed at the outer surface of the anode.

During SO-DCFC operations, oxygen in air is reduced to oxygen ions (O^{2-}) at the cathode and then dissolved into the electrolyte. The O^{2-} is then transported through the solid electrolyte to the anode region. These oxygen reduction and transport processes are the same as those in SOFC cathodes and electrolytes. After O^{2-} ions reach the anode, it will react with the fuel to complete the fuel oxidation process, producing electrical power. In the case where carbon is used as the fuel,

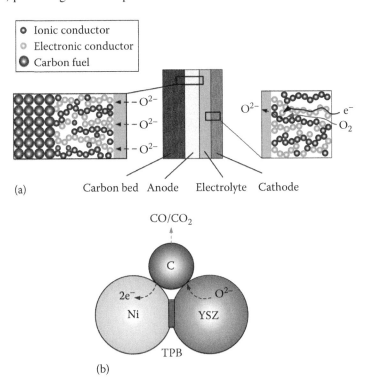

FIGURE 5.4 Sketch of (a) SO-DCFC working process and (b) ideal carbon oxidation mechanism.

the electrochemical reactions at the triple-phase boundary (TPB) of the gas ionic conductor (noted as YSZ) and electronic conductor (noted as metal Ni) can be expressed by Equations 5.1 and 5.7.

For SO-DCFCs, direct carbon oxidation of solid carbon fuels by oxygen ions rarely occurs because the poor solid-to-solid contact between the carbon fuel and the anode does not allow for both ion and electron transfers. From a fuel transport point of view, the pole diameter of the porous anode is usually less than 1 μm to maximize the volumetric densities of the interface between the sintered Ni particle and the YSZ particle [3]. The submicropores are also much smaller than the size of the carbon particles fed into the anode chamber (often with a diameter of 50–200 μm). Therefore, it is rare for carbon fuels to reach the Ni–YSZ TPB interface located deep inside the anode.

Considering the harsh nature of carbon oxidation at the anode of an SO-DCFC, various methods have been adopted to accelerate carbon conversion, including cracking carbonaceous fuels inside the anode to deposit carbon onto the Ni–YSZ interface, introducing molten media as an extra oxygen supplier or fuel carrier and gasifying solid carbons into gaseous fuels such as CO or syngas. More detailed carbon conversion mechanisms and techniques of carbon conversion will be discussed in the latter parts of this chapter.

5.3.2 Molten Media in DCFCs

Molten media such as molten hydroxide are being widely adopted as the electrolyte in DCFCs. Molten hydroxide electrolytes (made up of NaOH and KOH) possess advantages such as high ionic conductivity and high reactivity toward carbon fuels. Studies report that molten hydroxide DCFCs (MH-DCFC) can achieve peak power densities of 180 mW/cm^2, and average power density values of 40 mW/cm^2 over 540 hours, operating at 630°C [4]. Hackett et al. [5] tested various fuels in an MH-DCFC and achieved peak power densities of 84 mW/cm^2 for graphite-based carbon rods and 33 mW/cm^2 for coal-derived ones at temperatures of 675°C. The anode reaction can be expressed as Equation 5.18:

$$C + 4OH^- = CO_2 + 2H_2O + 4e^- \qquad (5.18)$$

The actual reaction taking place can be complex however, because the oxidation of carbon is a 4-electron transferring reaction with the products of oxygen partial reduction such as superoxide $\left(O_2^-\right)$ or peroxide $\left(O_2^{2-}\right)$ possibly playing a role during fuel cell operations.

An issue affecting MH-DCFC operations is carbonate formation through the reaction between CO_2 and OH^-, as expressed by Equation 5.19:

$$2OH^- + CO_2 = CO_3^{2-} + H_2O \qquad (5.19)$$

To resolve this, Zecevic et al. [4] proposed several methods to suppress this carbonation process, including increasing the water content in the electrolyte by feeding the cathode with humidified air and introducing extra oxyanions such as metal oxides, pyrophosphates, and persulfates into the melt system.

5.3.2.1   Molten Carbonate in DCFCs

Eutectic mixtures of molten carbonates (often binary or ternary mixture of Li_2CO_3, Na_2CO_3, and Li_2CO_3) are applied to DCFCs due to their ionic conductivities and tolerances toward produced CO_2. A DCFC using molten carbonates as the electrolyte is referred to as a molten carbonate DCFC (MC-DCFC).

In MC-DCFCs, porous ceramics are used as the liquid-form electrolyte supporter, while lithiated nickel oxide (NiO)-based materials are used as the cathode. During fuel cell operations, molten carbonates fill up pores of the ceramic supporter and build ionic-transporting pathways connecting the cathode and anode. Carbon slurries made of pulverized carbon and molten carbonates are introduced into the anode chamber and are oxidized by ions transported through the molten carbonate electrolyte to generate power. The reactions at both electrodes are expressed as Equations 5.20 and 5.21 [6]:

Anode:

$$C + 2CO_3^{2-} = 3CO_2 + 4e^-$$ (5.20)

Cathode:

$$O_2 + 2CO_2 + 4e^- = 2CO_3^{2-}$$ (5.21)

The anode reaction expressed by Equation 5.20 shows that both the carbon and carbonate anions $\left(CO_3^{2-}\right)$ are consumed during the carbon oxidation process. Therefore, to maintain CO_3^{2-} concentrations, CO_2 is continuously fed to the cathode chamber to compensate for the consumed CO_3^{2-} ions in Equation 5.20. Cherepy et al. [6] operated an MC-DCFC at 800°C using 32–68 mol.% Li-K eutectic molten carbonates fueled with different types of carbon fuels including activated carbon, biomass-derived char, carbon black derived from cracking, and petroleum coke. The obtained peak power densities were found to range from 40 to 100 mW/cm². This performance difference is closely related to the carbon microstructures, in which the crystalline parameters changed the chemical properties of the various carbon fuels in terms of their reaction activities.

The configuration of combining molten mediums with porous supporters requires the proper management of molten carbonates in the porous medium. This is because a lack of molten carbonate will result in a limited number of ionic transport pathways in the ceramic supporter, leading to unacceptably high ohmic resistances in DCFCs. However, too much carbonate is also detrimental because flooding and deactivation of the cathode may occur. To resolve this flooding issue, tilted electrolyte supporters were developed to prevent cathodes from totally flooding. In addition, a carbonate circulation system is also required to collect the molten carbonate that spills at the cathode side and return it to the anode side [7].

Compared with techniques using molten media as electrolytes, SOFCs using solid electrolytes are more stable from a structural point of view, because CO_2 circulation and carbonation between electrodes are not required in SOFCs. However, this advantage is compromised by poor contact conditions and slow mass transport rates. Therefore, a promising method for carbon conversion is to combine molten carbonates with solid-state electrolytes. This way, the dense and gas-tight, solid-state electrolyte can prevent the leakage of molten media, and molten carbonate can serve as a perfect medium for carbon conversion.

The combination of molten carbonate and solid-state electrolyte was demonstrated by Pointon et al. [8]. In their configuration, a YSZ blind tube was inserted into a molten carbonate bath with Pt serving as the in-tube cathode. This DCFC system operated under 665°C and 700°C, at a constant working voltage of 0.5 V. It was observed that the power density degraded sharply from 20 to 10 mW/cm^{-2} after nine hours of testing. The main reasons for this degradation were attributed to the formation of CO_2 bubbles at the YSZ–molten carbonate interface and the corrosion of YSZ in the molten carbonate. Results of etching test showed that the ceria-based electrolyte appeared stable in Li-K molten carbonate. The Na-K molten mixture and eutectic melt consisting of Na_2CO_3 and Li_2CO_3 also showed good compatibility with YSZ.

Nabae et al. [9] introduced a Li-K molten carbonate–based carbon slurry (at a molar ratio of carbon:carbonate = 1:1) into the anode chamber of a YSZ electrolyte supported bottom-type SOFC and achieved a peak power density of 13 mW/cm^2 at 900°C. The introduction of the carbonate was thought to be able to extend the electrochemical reactive sites from the solid-state anode to the carbonate melt, allowing for easier carbon fuel transport. The enhancement to fuel cell performances by molten carbonates is significant according to Kaklidis et al. [10] where in their experiment, introducing a Li-K molten carbonate–based carbon slurry (carbon:carbonate = 4:1, molar ratio) into the anode chamber of a YSZ-supported SOFC increased power densities from 15 to 25 mW/cm^2 at 800°C.

By using anode-supported SOFCs and high-performance cathodes made of LSC ($La_{0.6}Sr_{0.4}CoO_{3-\delta}$), Jiang et al. [11] demonstrated steady power densities of 140–200 mW/cm^2 during a 13-hour constant potential discharge test at 700°C. The YSZ electrolyte also showed good tolerance toward Li-K corrosion under inert and reducing atmospheres, and lithium zirconate formations (resulting in a less stable electrolyte) were only observed under oxidizing atmospheres.

There may be several different oxidation mechanism pathways for carbon fuels in molten carbonates. Jain et al. [12] believed that CO_3^{2-} is the reactive ion for carbon oxidation with the oxidation process of carbon by carbonate following Equation 5.20. They thought that the CO_3^{2-} consumed during fuel cell operations is regenerated by the chemical reaction between produced CO_2 and transported O^{2-} at the solid oxide electrolyte, as expressed by Equation 5.22:

$$CO_2 + O^{2-} \rightarrow CO_3^{2-} \tag{5.22}$$

The mechanisms of carbon oxidation involving oxygen ions were also proposed because there is sufficient O^{2-} being supplied to the anode region [13]. The only difference between the solid-state anode and the molten carbonate anode is the place where the oxidation reaction takes place. For solid-state anodes in SOFCs, carbon can only get access to O^{2-} by being transported to the TPB, whereas for molten carbonate anodes, O^{2-} can interact with carbon in the liquid-form anode. In both cases, O^{2-} is supplied by the ionic-conducting ceramics and the equilibrium shown in Equation 5.23:

$$CO_3^{2-} \leftrightarrow CO_2 + O^{2-} \tag{5.23}$$

Both oxidation routes may occur simultaneously during the fuel oxidation process because the existence of Equation 5.23 allows the eutectic mixture to act as both a CO_3^{2-} conductor as well as an oxygen ion conductor.

Apart from the corrosion effects of electrolytes, another issue for molten carbonate anodes is its poor electronic conductivity. Deleebeeck et al. [14] reported a strong dependence between current collection materials and fuel cell power densities. When Ni mesh is used as the current collector, a peak density of 30.6 mW/cm² can be reached in a 62–38 mol.% Li-K molten carbonate system at 800°C. Peak power densities of other current collectors are listed as follows: Pt mesh: 23.0 mW/cm²; Ag mesh: 19.8 mW/cm²; and Au mesh: 9.8 mW/cm². Deleebeeck et al. [14] attributed the differences in peak power densities to the different catalytic effects and chemical stabilities of the various metals. From photos of the various metal meshes taken before and after the electrochemical tests, significant changes in color from silver white to gray were found in the silver mesh, wires in the gold mesh became thinner, holes developed in the Ni mesh, and the structure of the Pt mesh changed greatly. These external appearance changes are potential indicators for physical property changes of the metal surfaces and, subsequently, changes to their electronic conductivity during exposure to the molten carbonate. Xu et al. [15] increased the power density of a molten carbonate anode from 40 to 120 mW/cm² by replacing a gold mesh two-dimensional current collector with a three-dimensional current collector (a spiral tube rolled with copper foil). Therefore, more conductive anodes are required for carbon conversions, and anodes made of metals in liquid form are needed for DCFCs.

5.3.2.2   Liquid Metal in DCFC

Metal Sn has been introduced into the anodes of SOFCs to suppress coking effects when fueled with carbon-containing fuels. Myung et al. [16] demonstrated this by added Sn to the Ni-GDC anode of a CH_4-fueled SOFC working at 650°C. The Sn-doped anode continuously demonstrated at a constant current density of 0.5 A/cm² for over 250 hours without degradation when dry CH_4 was used as a fuel, whereas the anode without Sn doping failed resulting from carbon deposition after 1.5 hours of operation under the same working conditions. This observation was further proved by Yang et al. [17]. In their experiment, the Sn-doped Ni-SDC anode produced better stabilities under a humidified CH_4 atmosphere at 700°C than the undoped one. Microscopic photos showed that highly reactive Sn–Ni intermetallics were formed, and these could serve as active reaction sites for converting carbon to provide the demonstrated fuel cell performances.

Liquid metal Sn is further suggested to be a suitable material for solid carbon fuel conversions. Ju et al. [18] introduced a mixture of Sn and carbon black into an electrolyte-supported SOFC using an Ni–YSZ anode, and peak power densities increased from 14 to 60 mW/cm² at 900°C. In their experiment, power densities further increased from 20 to 200 mW/cm² when fueled with CO under the same conditions and configurations, demonstrating that Sn metal also has the capability of converting gaseous intermediate products of carbon oxidation into power. In addition, Sn metal in its liquid form can act as a dispersion agent of carbon particles, carrying carbon fuels into the pores of Ni–YSZ anodes. It can also accelerate the carbon oxidation process by improving the contact between carbon fuels and solid-state surfaces in the anode region and catalyze CO produced during the carbon oxidation process. Therefore, these liquid-Sn-promoted enhancements to carbon and CO oxidation processes significantly improve DCFC performances.

Another type of SO-DCFC directly employ liquid metal as the anode where the solid-state anode is replaced by a liquid metal. This concept was demonstrated by Tao et al. [19] and Jayakumar et al. [20] separately. By using a liquid metal anode, the metal is first electrochemically oxidized to form a metal oxide and is then reduced by a carbon fuel, as expressed by Equations 5.24 and 5.25:

$$M + nO^{2-} \rightarrow MO_n + 2ne^- \tag{5.24}$$

$$MO_n + \frac{n}{2}C \rightarrow M + \frac{n}{2}CO_2 \tag{5.25}$$

The sketch of a liquid metal anode DCFC is shown in Figure 5.5.

Jayakumar et al. [21] fabricated a DCFC using liquid Sb as the anode, with ScSZ (Sc-stabilized zirconia) as the supporting electrolyte and LSF-ScSZ ($La_{0.8}Sr_{0.2}FeO_3$) as the cathode. Their fabricated DCFC exhibited a stable performance of 350 mW/m^2 at 700°C. The density of Sb_2O_3 is less than that of metal Sb, allowing it to float on top of the liquid Sb surface, leading to a separate reactor. When Sb oxides are reduced by carbon fuels, metallic Sb can sink and flow back to the bottom part of the electrochemical reactor. This circulation of Sb and Sb_2O_3 enables the fuel cell to operate continuously on solid carbon fuels and avoid influences of ash on fuel cell performances. A sketch of this DCFC system is demonstrated in Figure 5.6.

If the supply of carbon fuels is insufficient however, the fuel cell system will continue to generate power by consuming the chemical energy stored in metallic Sb, causing degradation.

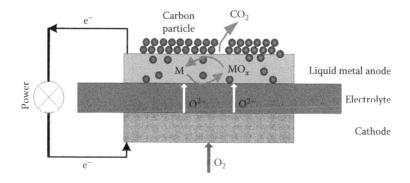

FIGURE 5.5 Sketch of a liquid metal anode DCFC.

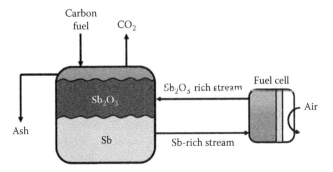

FIGURE 5.6 Demonstration of a liquid Sb anode DCFC with a reduction cycle. (From Jayakumar, A. et al., A direct carbon fuel cell with a molten antimony anode, *Energy Environ. Sci.*, 4, 4133–4137, 2011. Reproduced by permission of The Royal Society of Chemistry.)

There are several challenges to liquid metal anodes. One of which is the blocking effect of metal oxides, in which the melting points of many metal oxides are much higher than their elemental metal forms and are even higher than that of DCFC operating temperatures, resulting in the blocking of reaction sites. For example, the melting point of Sn metal is 232°C, but the melting point of its oxidized form such as SnO_2 is as high as 1630°C, which exceeds the operating temperatures of a DCFC. Although Sn has shown fascinating carbon conversion capabilities as described previously, solid, insulating SnO_2 films will deposit on anode–electrolyte interfaces if large currents are extracted from the anode. This impedes further electrochemical reactions, resulting in sudden drops of fuel cell power outputs. Nonetheless, Sn is still considered to be a promising metal for liquid metal carbon conversion anodes in terms of its vapor pressure at high temperatures, especially when compared to other metals such as Sb. The vapor pressure of Sn is 9.53×10^{-5} Pa at 800°C, whereas Sb has a vapor pressure of 20.2 Pa at the same temperature. This smaller vapor pressure value means a much slower escape rate for Sn than its Sb component. From a thermodynamic point of view, the OCV value of liquid Sn anodes versus air is measured to be 0.9 V at 800°C, which is higher than that of Sb anodes versus air (around 0.7 V at 800°C). For Sn liquid anodes, efforts need to be taken to prevent liquid Sn anodes from forming SnO_2 insulating films. To achieve this, strong, penetrable reducing agents such as H_2 [22] or proper selections and supplements of carbon fuels [23] are needed to reduce the SnO_2 film. Other technical issues such as liquid metal or metal oxide losses due to evaporation, influence of ash contents in carbon fuels, and fuel cell stabilities need to be resolved as well.

Apart from reactive anodes, "inert" liquid metal anodes are also being developed based on Ag. Liquid Ag can dissolve substantial amounts of oxygen (4.4×10^{-8} mol/cm^3 at 1000°C) [24], which can provide a proper oxygen source for carbon oxidation. Javadekar et al. [25] used a liquid silver anode as an oxygen reservoir to oxidize carbon fuels in their tested DCFC. In this configuration, an Ag-based carbon slurry is fed to the anode chamber, with liquid silver acting as a carbon dispersion agent an oxygen bank and a continuous electron collector. The liquid silver in the anode chamber maintains its metallic state during DCFC operations because the oxide of silver, Ag_2O, is thermodynamically unstable under DCFC operation temperatures. Thus, the liquid silver anode is regarded as "inert." The tested DCFC with the liquid Ag anode operated at 1273 K, with an OCV value of 1.12 V. The impedance of this liquid Ag anode was as high as ~100 Ω cm^2, resulting from the slow rate of oxygen transport inside the liquid Ag. An analysis of the liquid Ag anode revealed that the transport of oxygen in the anode region to oxidize carbon fuels is critical for the utilization of liquid Ag as a carbon conversion anode in DCFCs.

Molten media in SO-DCFCs improve contact conditions between carbon and anodes and speed up carbon transport. With the help of highly reactive molten carbonates and liquid metals, transport and conversion processes of carbon become less sluggish. However, because these high-temperature fuel cell systems consist of liquid-phase components, stability and corrosion of molten media are still challenges. Transport of reactive species and selection of proper molten media with both high efficiency and acceptable performance still require intensive research.

5.4 ROLE OF CO IN CARBON CONVERSION

In DCFCs, pulverized solid carbon fuels are often supplied directly to the anode chamber. Direct reactions between the solid carbon and oxygen ions rarely take place in conventional Ni–YSZ anode systems however, because the ionic conductivity of YSZ is much smaller than the electronic conductivity of Ni. Therefore, electrochemical reactions at the anode often take place in the electrolyte adjacent regions. When solid carbon is supplied as the fuel, the diameter of the carbon particles (usually hundreds of micrometers) is much larger than that of the anode pores (often of several micrometers), making the transport of carbon particles through the porous anode to the electrochemical reactive interface impossible. Kulkarni et al. [26] utilized a mixed ionic electronic conductor (MIEC) as a carbon conversion electrode to see if the ionic conducting anodes can supply sufficient oxygen ions to oxidize carbon fuels allowing direct oxidation to take place at the outer surface of the anode where the carbon is in contact with the anode. However, in their experiment, the reaction between the solid carbon and the solid-state anode was found to be limited by poor contacting conditions between the solid phases. This was true even in the case of the deposited carbon at the anode, where the contact conditions between the carbon and electrode should be much better, carbon oxidation still suffered from high activation polarizations [27]. Despite this, solid carbon fuels consumed in SO-DCFCs with solid-state anodes are often converted to CO, which is a gaseous fuel with much higher diffusivity than solid carbons. Therefore, CO plays an important role in solid carbon oxidation at the anodes of DCFCs. This section will discuss the role of CO in DCFCs and review several methods to promote CO production at the anode.

5.4.1 "CO SHUTTLE" MECHANISM

CO production during solid carbon oxidation at the DCFC anode is observable if carbon black is used as the fuel, because carbon black is a good absorbent for O_2, H_2O, and other oxygenated species. These adsorbed species then trigger carbon fuel gasification reactions at elevated temperatures. Furthermore, gaseous oxygen pumped to the anode through the electrolyte can also oxidize carbon fuels to produce CO or CO_2 [28]. For DCFCs operating at intermediate temperatures as low as 600°C, Boudouard reaction can still produce CO.

Once the gaseous species (CO or CO_2) are introduced into the DCFC anode chamber, the reactions are initiated, and the continuous operation of the DCFC through a "CO shuttle" mechanism as proposed by Gur [2]. This proposed "CO shuttle" mechanism is actually a CO-powered SOFC reaction process with carbon presence in the anode chamber, in which the CO formed inside the anode chamber will diffuse into the porous anode and be electrochemically oxidized. The CO_2 produced by the electrochemical oxidation of CO will subsequently react with carbon to produce CO via the Boudouard reaction. This CO–CO_2 cycle can continuously power the DCFC as long as carbon fuel is supplied. The different DCFC performances under various Ar flow rates at the anode are evidence for this "CO shuttle" mechanism. Li et al. [27] carried out DCFC tests under different Ar flow rates through the anode in the presence of solid carbon physical contacting the DCFC anode. Fuel cell performances under smaller Ar flow rates were found to be better than that under larger Ar flow rates. This phenomenon suggests that DCFCs are mainly powered by the electrochemical oxidation of CO because under smaller Ar flow rates, the CO generated by the Boudouard reaction accumulates in the anode and its adjacent regions, leading to increased performances, whereas under larger Ar flow rates, the CO produced is diluted and quickly blown away by the Ar flow, leading to decreased DCFC performances.

Reaction mechanisms between carbon and CO_2 have been systematically studied [29,30]. A mechanism for the C–CO–CO_2 reaction system of DCFC operations was presented by Lee et al. [31], in which two types of carbon sites were considered due to the complex nature of the carbon materials. In their presented mechanism, C_f denotes the free carbon site available for adsorption, and C_b denotes the exposed bulk carbon site when a free carbon site is removed from the solid carbon fuel. Adsorbed species, such as oxygen adsorbed on carbon (O(C)) and CO adsorbed on carbon (CO(C)), were also taken into consideration. CO production through carbon gasification in their presented reaction mechanisms is expressed by Equations 5.26 through 5.28:

$$CO_2 + C_f \Leftrightarrow CO + O(C) \tag{5.26}$$

$$C_b + O(C) \rightarrow CO + C_f \tag{5.27}$$

$$C_b + CO_2 + O(C) \rightarrow 2CO + O(C) \tag{5.28}$$

The generation of CO_2 through the consumption of CO was also considered in their mechanism, as expressed by Equations 5.29 and 5.30:

$$C_f + CO \Leftrightarrow CO(C) \tag{5.29}$$

$$CO + CO(C) \rightarrow CO_2 + 2C_f \tag{5.30}$$

Normally, carbon gasification is a process in which heterogeneous reactions take place on material surfaces. Closed pores buried under carbon surfaces can open when existing open pores merge due to fuel consumption. Random pore models of porous material gasification can be employed to evaluate the specific surface areas (S_{gC}, with a unit of area over mass, i.e., m²/g) of carbon during reactions, as expressed in Equation 5.31:

$$S_{gC} = S_{gC,0}\sqrt{1 - \psi \ln\left(1 - x_C\right)} \tag{5.31}$$

where

$S_{gC,0}$ is the initial value of the specific surface area

x_C is the carbon conversion ratio (ratio between carbon conversion rate integrated over time and initial mole of carbon applied)

ψ is the structural parameter determined by fitting to measured specific surface areas

Direct contact between carbon fuels and DCFC anode surfaces can provide extra thermodynamic benefits to the "CO shuttle" mechanism, such as heat generation during fuel cell operations (heat released during electrochemical reactions or Ohmic heat). This extra heat can be removed to maintain optimal fuel cell operation temperatures and be used to warm up carbon fuels and maintain CO productions due to the Boudouard reaction being endothermic. This coupling of electrochemical reactions with carbon gasification, or in simpler terms, heat transfer from fuel cell to solid carbon, improves efficiencies and reduces DCFC system complexities.

Discussions earlier suggest that the performance of a DCFC with solid carbon fuels fed directly into the porous anode mainly come from the oxidation of CO during carbon gasification and rarely from the oxidization of carbon using oxygen ions (O^{2-}). Therefore, accelerating the rate of CO generation is a promising method to improve DCFC performances. Because the Boudouard reaction is endothermic, it will benefit from increased fuel cell operation temperatures, which will move the direction of the equilibrium between CO and CO_2 forward, increasing CO generation. This, along with sufficient fuel supplies, improved ionic conductivities, and more active reaction mechanisms at higher operation temperatures will increase the power density of DCFCs operating at elevated temperatures.

5.4.2 Steam Gasification

The introduction of CO_2 or H_2O (steam) into the anode chamber as a gasification agent can increase the amount of CO/H_2 fuels (syngas). The desired reaction between steam and carbon is expressed by Equation 5.32:

$$C + H_2O = H_2 + CO \tag{5.32}$$

When provided with both carbon and CO, the Boudouard reaction and its reverse reaction are expressed by Equation 5.33:

$$C + CO_2 \Leftrightarrow 2CO \qquad (5.33)$$

Water–gas shift reactions can also take place if excessive steam is supplied to the anode, as shown by Equation 5.34:

$$CO + H_2O = H_2 + CO_2 \qquad (5.34)$$

Both CO-producing reactions are endothermic, with the overall enthalpy change of Equation 5.32 being +135.7 kJ/mol at 800°C, and +174.6 kJ/mol for Equation 5.33 at the same temperature. The water–gas shift reaction shown in Equation 5.34 is a moderately exothermic reaction however, with an enthalpy change of −34.1 kJ/mol at 800°C. Therefore, to achieve satisfactory equilibriums, gasification reactions (Equations 5.32 and 5.33) are often arranged separately from shifting reactions (Equation 5.34). Because gasification processes require high temperatures, pure oxygen is often introduced into the gasifier to burn parts of the coal for heat production (N_2 is often excluded from the reaction system to prevent the dilution of product gases and to overcome the difficulties of carbon capture). The H_2 and CO produced during gasification and shifting processes, regardless of their derivation, are transported to the fuel cell and consumed to produce power. Compared with dry gasification (gasifying carbon with CO_2), steam gasification for DCFCs is unpopular because it requires more preheating and extra water input (at least at start-up working conditions). However, steam gasification was found to be six to eight times faster than dry gasification [32,33] and the H_2 rich fuel gases produced can be oxidized more rapidly by the anode, leading to better fuel cell performances.

Zhou et al. [34] developed a cathode-supported tubular DCFC using an Ni–ScSZ anode. Steam was injected into a stand-alone carbon bed to produce H_2-rich fuel gases, and the peak power density of their steam-driven fuel cell was 91.1 mW/cm^2 at 850°C and 172.7 mW/cm^2 at 900°C. This system was further optimized by Deng et al [35]. In their configuration, carbon black loaded with catalysts (catalysts took up 1 wt.% of carbon) was enclosed into the inner space of an anode-supported fuel cell. Steam was sufficiently supplied by bubbling carrier gases (N_2) through a hot water bath kept at 70°C. Among all the catalysts tested (CeO_2, K_2O, and CaO) for steam gasification, CeO_2 showed the best performance with a peak power density of 214 mW/cm^2 at operating temperatures under 850°C. It was observed that excessive amounts of water introduced into the anode chambers can oxidize Ni metals inside the anodes at 750°C and that the thermal effects of endothermic steam gasification can affect fuel cell temperatures.

Several systems integrating gasification processes and fuel cells (IGFCs) have been proposed as a clean coal utilization technology [36,37]. Steam gasification allows for both pulverized coal and water–coal slurries to be used as feedstocks, leading to wider selections of gasifiers. There are several issues requiring consideration for steam-gasification-driven SOFCs however. One issue is that large amounts of water must be supplied to the fuel cell system because the production of gaseous fuels requires steam as the gasification agent or shifting reactant. Here, the circulation of the anode flue gas can fulfill parts of the water demand. Another issue is the high H_2 content, which can form volatile compounds with the elements in coal under reducing atmospheres. Hydrides such as H_2S, NH_3, PH_3, and AsH_3 are potential threats to anode catalysts. Therefore, carefully designed fuel gas purification processes are necessary.

5.1.3 CATALYTIC GASIFICATION

Catalysts can be introduced into DCFCs to promote carbon conversions. Ceria-based ceramics [38,39] and perovskite type of materials [40] have been employed as catalytic anodes for carbon conversion. However, because the rate-determining step during the operation of these types of DCFCs is the carbon gasification to produce CO, much higher temperatures and pressures are needed. To accelerate CO production, catalysts are normally loaded with carbon and introduced into DCFC anodes. Of these, iron-based [41,42] and alkaline-based [43,44] catalysts loaded into solid fuels have been extensively studied.

Tang et al. [45] fabricated an electrolyte-supported tubular DCFC with a GDC–Ag mixed anode and used an activated carbon loaded with Fe-based catalysts at a ratio of C:Fe = 4:1 (mass ratio) as the fuel. By using this configuration, peak power densities at 800°C increased from 24 to 45 mW/cm². The authors attributed this improvement to accelerated CO production rates. The results of these gasification tests carried out at given temperatures show an increase in CO content in the product gases in the presence of Fe catalysts between 700°C and 850°C. Here, the CO is produced by the valence change of Fe elements, as shown by Equations 5.35 and 5.36:

$$Fe_mO_n + CO_2 \rightarrow Fe_mO_{n+1} + CO \qquad (5.35)$$

$$Fe_mO_{n+1} + C \rightarrow Fe_mO_n + CO \qquad (5.36)$$

Li et al. [46] loaded commercial carbon black with a K-based catalyst at a ratio of C:K = 10:1 (mass ratio). The temperature of the carbon bed and anode-supported DCFC (employing Ni–YSZ as the anode) were controlled separately, with carbon black temperatures varying between 700°C and 1000°C at intervals of 50°C, and fuel cell temperatures being kept at 750°C. It was observed that a peak power density of 185 mW/cm² can be reached when the K-loaded carbon black was heated to 850°C. This is five times higher than that of the DCFC without catalysts working under the same conditions. In situ gas analysis performed by gas chromatography at the various working conditions shows that CO took up 40 mol.% of the product gas flow when CO_2 was introduced into the anode chamber as a gasification agent at a flow rate of 50 sccm. A detailed catalytic gasification mechanism was proposed and validated by Yu et al. [47], in which the K-based catalytic gasification mechanism of carbon involves K_xO_y as potassium-rich clusters in the K-based catalyst ([KO]), oxygen complexes "dissolving" into the potassium-rich clusters (O[KO]), and absorbed O[KO] clusters on the carbon surface (O[KO](C)). Here, the carbon atoms are classified into two types: free carbon atoms (C_f, a carbon atom available for adsorption) and bulk carbon atoms (C_b, an underlying atom that can be exposed if a free carbon site is desorbed from the solid carbon fuel). CO can be produced by oxygen dissolution in the potassium-rich clusters, as shown by Equation 5.37:

$$CO_2 + [KO] \rightarrow CO + O[KO] \tag{5.37}$$

The O[KO] complex made up of dissolved oxygen and potassium-rich clusters can then react with free carbon atoms, as shown by Equation 5.38.

$$C_f + O[KO] \rightarrow O[KO](C) \tag{5.38}$$

The carbon–oxygen–potassium-containing complex O[KO](C) can then turn a bulk carbon atom (C_b) into a reactive one (C_f) and generate CO, as shown by Equation 5.39:

$$C_b + O[KO](C) \rightarrow CO + C_f + [KO] \tag{5.39}$$

According to the calculated results of gasification models built based on the mechanism presented in Equations 5.37 through 5.39, CO production rates of catalytic gasification are two orders of magnitude faster than that of Boudouard reactions without catalysts under the same operation temperatures, demonstrating the necessity of catalytic gasification.

Wu et al. [48] developed a complex catalyst composed of Fe_2O_3, Li_2O, K_2O, and CaO, to take advantage of both subgroup metals and alkaline/alkaline earth metals. The catalyst was then introduced into different types of carbon black and graphite fuels. The resulting promotional effects of CO production rates induced by this catalyst showed strong dependence on both the carbon bed temperature and the carbon specific surface area. A peak power density of 297 mW/cm^2 was observed on an Ni–ScSZ anode-supported SOFC operating under 850°C. SEM photographs showed that the "CO shuttle" mechanism, which requires the CO_2 formed at the TPB to be diffused out of the anode, can result in high CO_2 concentrations at the anode and its adjacent regions, effectively suppressing carbon deposition at the Ni-based anode. Carbon black loaded with complex catalysts was also introduced into a tubular SOFC as fuel by Yang et al. [49]. They suggested that minimizing CO loss from the anode chamber is key to promoting fuel efficiencies, and proposed a porous ceramic matrix (made of SDC) soaked with molten carbonate (ternary mixture of Li_2CO_3, Na_2CO_3, and K_2CO_3) to be mounted to the opening of the tubular fuel cell as a CO_2 permeable membrane. This membrane, made up of both oxygen ions (O^{2-}) and CO_3^{2-} conductors, is a selective membrane only permeable to CO_2, sealing generated CO into the anode chamber while emitting CO_2 as the flue gas. The installation of such a membrane increases the difficulty of continuous refueling in the fuel cell however. The researchers regarded this type of electrochemical device as a "carbon–air" battery, but this device can still be considered as a DCFC because the reaction mechanisms in this "carbon–air" battery are the same as those in SO-DCFCs. With the help of the selective membrane and the complex catalyst, high CO concentrations can be achieved with the tested Ni–YSZ anode-supported fuel cell. The peak power density of the cell reached 279.3 mW/cm^2 when operated at 850°C. When the discharge of the DCFC was complete, unreacted carbon was still present in anode chamber, with only 14.36% of the carbon loaded being converted into power. The researchers attributed the carbon residue to catalyst deactivation through sintering and agglomeration when the DCFC was operated at high temperatures. Zhong et al. [50] introduced Al$_2$O$_3$ as a sintering inhibitor for Fe$_2$O$_3$–K$_2$O complex catalysts using the same tubular DCFC testing configuration as Yang et al. In their experiment, peak power densities reached 292 mW/cm^2 with catalyst sintering largely avoided and high fuel conversion ratios of 98.7% achieved.

5.4.4 INDIRECT CARBON FUEL CELL

As discussed earlier, DCFCs with carbon deposited inside the anode or carbon fuels introduced directly into the anode chamber demonstrate strong coupling between the fuel cell and the carbon. The gasification products of the carbon bed can power the fuel cells while the oxidation products such as CO$_2$ and H$_2$O can diffuse back to serve as gasification agents. The heat released by the electrochemical reactions can warm up the carbon bed, allowing endothermic reactions to take place. Due to the direct contact between the fuel cell anode and the carbon for fuel oxidation, these fuel cell configurations are regarded as the "direct" conversion of carbon.

Although theoretical efficiencies of these DCFCs are incredibly high, severe challenges induced by solid carbon fuels still exist. Known-structured/composited and purified carbon black, activated carbon, or graphite are mostly used for example fuels in DCFC testing. However, real-world carbon materials such as the various types of coals, chars, biomasses, or municipal wastes are much more complex, in which poisonous elements can form gaseous species during carbon gasification and migrate from the fuel to the anode and deactivate the anodic catalyst. The coupling of exothermic electrochemical and endothermic gasification reactions requires both thermal conduction of solid phases and heat convection of gas flows, causing a cross-sectional thermal gratitude in fuel cell anodes. This gratitude will grow steeper during load-shifting working conditions. To accelerate carbon gasification rates under fuel cell working temperatures and pressures, catalytic gasification processes are necessary. However, the reclamation of catalysts loaded onto coal (as mentioned in previous studies) is challenging and needs to be solved by either novel loading methods or development of cheaper catalysts. Several studies have claimed that even when fabricated DCFC stacks were operationally feasible [49,51], the continuous feeding of carbon in scaled-up DCFC systems was difficult. For example, a promising fueling method was proposed by Lee et al. [52,53], in which carbon materials were supplied by a fluidized bed to a DCFC, but attrition of the anode by the fluidized carbon fuel was an issue.

Because most challenges are caused by carbon conversion, other technical methods such as separating the carbon from the anode or even carrying out the gasification reaction in stand-alone gasifiers were proposed as possible ways to drive fuel cells. And although the product gases of these methods are still injected into the anode chamber, the energy transferred between the carbon and the fuel cell is largely weakened by enlarged distances or separated reactors. Several batch experiments were carried out with respective temperature controls of carbon beds and fuel cells [28,46]. This option is referred to as "indirect carbon fuel cell (ICFC)." The concepts of "direct" or "indirect" in this section are only related to the connection between the carbon and the fuel cell, with the term "indirect" representing a less coupled situation between carbon gasification and power generation. A novel ICFC system concept was proposed by Ong et al. [54], in which carbon fuels were gasified in a gasifier with the heat released by electrochemical reactions being rejected to the gasifier (rejection of heat is accomplished by advanced high-temperature heat exchangers, e.g., high-temperature heat pipes [55]). This ICFC system demonstrated enhanced energy exchanges between the reactors and the advantage of separating carbon from anodes.

ICFC systems have more flexibility in terms of design by leaving the difficulties of carbon conversion to gasifiers and several hybrid systems combining coal gasifiers, fuel cell stacks, and other bottom cycles have been proposed [36,56]. For example, in a "FutureGen" power plant, a steam gasifier was integrated with an SOFC stack, gas turbine, and steam turbine. A large-scale expedient testing of an atmospheric direct fuel cell/turbine hybrid system was explored by the National Energy Technology Laboratory (NETL) and FuelCell Energy Inc. using a 220 kW SOFC stack combined with an initial Capstone 30 kW and a 60 kW modified microturbine generator. This hybrid power system operated for over 6800 hours in total, achieving efficiencies of 52%, which is incredibly high even for natural gas systems.

NETL also conducted several system studies on IGFC power plants [57,58] for centralized power generation with near-zero pollutant emissions and carbon capture. IGFC power plants are analogous to an integrated gasification combined cycle (IGCC) power plant, but with gas turbine power islands replaced by SOFC power islands. The schematic demonstration of this proposed coal-based hybrid cycle power plant is shown in Figure 5.7, in which a ConocoPhilips E-Gas™ gasifier (CoP) and an SOFC stack operating under atmospheric pressures are adopted as the baseline of the IGFC plant. The syngas generated here goes through a purification step before

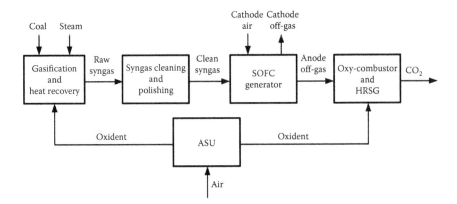

FIGURE 5.7 IGFC system configuration. (From Newby, R. and Keairns, D., *Analysis of Integrated Gasification Fuel Cell Plant Configurations*, February 22, 2011, National Energy Technology Laboratory, USA. Copyright 2009 DOE/NETL.)

being introduced into the SOFC stack. The pressurized vessels are connected to the SOFC stack via a syngas expander to recover part of the energy.

Other advanced coal-based technologies, such as supercritical pulverized coal (PC)-fired power plants and IGCC power plants, have also been investigated in other reports by NETL [59,60]. All cases of coal-based power plants were evaluated under the same ambient conditions (atmospheric pressure, dry bulb temperature, wet bulb temperature, etc.), fueled with the same type of coal (Illinois No. 6 coal) and with the net capacity of 550 MW. For the baseline case of IGFC power plants with CO_2 capture, the total plant cost (TPC, mainly includes equipment and real estate) was 2964 (2011$) kW^{-1} (2011 U.S. dollar per kilowatt) and the total overnight cost (TOC, mainly includes rail spur cost, switch yard cost, labor cost during plant construction, and cost of spare parts for plant operations added to TPC) was 3631 (2011$) kW^{-1}. For IGCC power plants without CO_2 capture, the TPC value was 2595 (2011$) kW^{-1} (original unit was (2007$) kW^{-1} in Ref. [59], an inflation rate of 8.5% was considered during dollar conversion, and following data concerning IGCC plants were processed in the same method) and the TOC was 3221 (2011$) kW^{-1}, being much cheaper than IGFCs. When CO_2 capture is considered however, investments for IGCCs sharply increase, with the TPC value becoming 3579 (2011$) kW^{-1} and the TOC value becoming 4433 (2011$) kW^{-1}. The dramatic increase in investment costs is also true for supercritical PC plants including CO_2 capture: TPC value increasing from 2026 to 3524 (2011$) kW^{-1}, and TOC increasing from 2507 to 4333 (2011$) kW^{-1}. Therefore, based on the current state of technology, combining IGFC systems with carbon capture technologies is financially attractive when CO_2 emissions are strictly regulated.

Due to increasing climate change in recent years, control of greenhouse gas emissions has become a heated topic around the world. Governments of developing and developed countries are facing similar dilemmas between choosing nationwide "carbon footprint" reductions or domestic economy developments. Therefore, technologies with less CO_2 emission and easy CO_2 capture, such as IGFCs, are promising solutions to tackle both energy demands and environmental issues. Although the first-year cost of electricity (COE) in IGFC baseline cases (170 (2011$) MWh^{-1}) is higher than that of IGCCs (153.5 (2011$) MWh^{-1} with carbon capture) and supercritical PC plants (142.8 (2011$) MWh^{-1} with carbon capture), several technical optimization procedures, including advanced gasifiers, long-life SOFC stacks, pressurized SOFCs, and high-efficiency inverters have been proposed and can decrease costs.

Analysis of IGFC systems carried out by NETL was based on large systems with net capacities of 500 MW. These are impossible to build with current fuel cell technologies. However, this half-imaginary analysis provides opportunities for IGFC systems to be compared directly with other advanced coal-fired power technologies at the same level. Easy capture of CO_2, enrichment of by-products (sulfur), less demand for water, higher net efficiencies over combustion, and merits of coal-based fuel cell systems have all been well demonstrated for IGFC systems, and direct comparisons have provided detailed capital costs as well as impressive efficiencies. These properties will potentially draw attention from both governments and investors, spurring the research and development of carbon-based fuel cells.

5.5   CARBON CONVERSION IN MOLTEN MEDIA

Molten media such as molten carbonates and liquid metals are often employed as carbon conversion anodes for SO-DCFCs. The carbon conversion processes in molten media are different from those in solid-state anodes. The carbon conversion processes in molten carbonate fuel cells (MCFCs) have been intensively studied. Conclusions from previous studies can also be adopted directly due to the similarities of the molten carbonate anodes of both molten carbonate DCFCs and SO-DCFCs.

5.5.1   WETTING CONDITIONS OF CARBON BY MOLTEN CARBONATE

The wetting properties of molten carbonates are important for MC-DCFCs because they dominate the formation of reactive areas in electrodes, affecting fuel cell performances. Reactive areas formed between molten carbonates and solid particles as a function of contacting angles are studied by Yoshikawa et al. [61], as shown in Figure 5.8.

The wetting of solid carbon fuels by molten carbonates is important in MC-DCFCs. The initial step of carbon conversion is to establish the electrical double layer between the molten carbonate and the solid carbon particles. This is crucial for both mass and charge transfer during electrochemical oxidation of carbon. Limited wettability between carbon and molten carbonate was reported by Peelen et al. [62] in which a graphite rod was inserted into a molten carbonate (62–38 mol.% Li-K carbonate) bath as a rigid sacrificial working electrode. Insufficient wettability of the carbon was observed based on visual inspections. Graphite has a typical laminar texture and its external surface has few lattice defects or active edge carbons necessary for

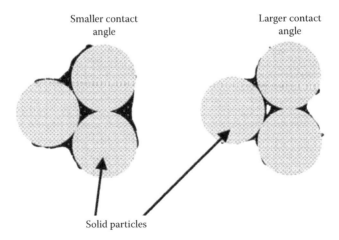

FIGURE 5.8   Wetting conditions of different contact angles (smaller on the left and larger on the right, dark zones as molten carbonate). (Reprinted from *J. Power Sources*, 158, Yoshikawa, M. et al., Experimental determination of effective surface area and conductivities in the porous anode of molten carbonate fuel cell, 94–102, Copyright 2006, with permission from Elsevier.)

anodic reactions. The lack of hydrophilic surface functional groups on the carbon surface makes graphite less attractive for ionic liquids such as molten salt. Therefore, introducing hydrophilic surface functional groups can help promote both the reactivity and wettability of carbon fuels toward molten carbonate. Surface modification of carbon fuels was carried out by Li et al. [63,64], in which various types of carbon samples were treated by acid washing (HCl, HNO_3, and HF) or air plasma. Lattice parameters (measure by XRD) and specific surface areas of the treated carbon samples did not change dramatically when compared with original samples. The treated carbon samples (particles) were then mixed with a 32-34-34 mol.% Li-Na-K eutectic carbonate melt. The surface-treated samples showed better electrochemical performances than the untreated samples. Taking the HNO_3 washed sample as an example, the power density of the original carbon was 53 mW/cm^2 at 700°C, whereas the HNO_3 washed sample under the same working conditions increased the power density to 74 mW/cm^2. The reason for this improved power density can be attributed to the increase of oxygen-containing surface functional groups induced by the acid wash. This change in surface functional groups by acid washing was also observed by an earlier study [65]. As observed, carbon washed by aqueous solutions of various acids brings substantial variations in chemical properties but leaves physical properties largely unchanged. HNO_3 treatment can generate large amounts of surface functional groups such as carbonyl (C=O) and carboxyl (–COOH), and HCl treatment can result in increased single-bonded oxygen functional groups such as phenols (–OH bonded directly to an aromatic hydrocarbon group), ethers (–O–), and lactones.

The time dependence of wettability between carbon materials and molten carbonates was observed by Chen et al. [66]. In their study, a graphite rod connected to gold wires was immersed in 62–38 mol.% Li-K mixed molten carbonate and used as a working electrode. A digital camera was attached to the testing configuration and optical observations were carried out simultaneously with electrochemical testing (fuel cell was operated at around 650°C). Due to the progressive wetting of the carbon electrode, the OCV of the DCFC, starting as low as 0.5 V, quickly increased to 0.67 V in 1.5 hours. The initial value of 0.5 V indicates low CO partial pressures or inert carbon fuels around the TPB. Once the carbonate was in contact with the graphite rod however, CO production by Boudouard reactions was triggered and oxygen ions (O^{2-}) in the melt were quickly adsorbed by carbon. The adsorption of O^{2-} can dramatically change the wettability of carbon surfaces, and the transition of a nonwetting meniscus to a wetting one makes it easier for capillary-effect-driven carbonate melt to creep into the pores of the graphite rod. In this process, the carbonate serves as an ionic-conducting phase and the graphite rod serves as both the electronic-conducting phase and the fuel. The TPB in this electrochemical system was established through the wetting of the carbon fuels by the electrolyte, at which carbon and CO are oxidized, resulting in increased OCVs of the DCFC. A similar dependence of OCV on immersion times was observed under the same testing conditions by Cooper and Selman [67], in which roughly 25 hours were needed for their OCV to reach a quasi-steady-state operating at 750°C. This time-consuming wetting process was also observed by Peelen et al. [62] in their AC impedance (EIS) measurements in which the charge transfer impedance of their test was observed to decrease to one-tenth of the initial value (from 35 to 3 Ω/cm^2) after the carbon anode was immersed in molten carbonate at 750°C for 24 hours. This change in charge transfer impedance demonstrates enhanced wettability after 24 hours of immersion.

The time-consuming soaking in molten carbonates can be accelerated by techniques such as elevating temperatures (as discussed earlier), mixing carbonates with rigid carbon anodes, and presoaking carbon fuels. Chen et al. [68] fabricated a composite anode made of needle coke and die-pressed carbonate and observed an increase in peak power density in their DCFC from 141 to 187 mW/cm^2 in a 62–38 mol.% Li-K carbonate system operating under 650°C. Composite anodes have made it easier for molten carbonates to wet the internal pore walls of rigid carbon anodes, rather than wetting just the external surfaces of pure carbon pellets. A similar fuel-pretreating method was carried out by Jain et al. [69]. In their study, an almost electrochemically inert pyrolyzed carbon, showing an OCV of 0.5 V at 750°C in 62–38 mol.% Li-K carbonate melt, was infiltrated by a saturated aqueous solution of Li-K carbonate. The OCV after infiltration increased to 0.63 V at the same temperature and further increased to 1.0 V after soaking the fuel in carbonate melt before feeding it to the fuel cell.

When dealing with real coal, contaminates (ash content) can also change the wettability of carbon fuels. Tulloch et al. [70] examined the effects of ash content on the electrochemical reactivity of carbon by loading minerals typically found in coal ash to graphite. Uncontaminated graphite samples were pressed and sintered into graphite pellets. The contaminated samples were prepared in a similar manner with graphite mixed in homogeneously in certain ratios. Electrochemical performances of different samples were tested by immersing the pellets into a 43.5-31.5-25 mol.% Li-Na-K carbonate mixture. Test results showed that the addition of clay materials, such as kaolinite and montmorillonite, can catalyze the oxidation of carbon fuels. Metal oxides and sulfides, such as anatase, alumina, and pyrite, can also slightly increase the carbon oxidation current. The addition of SiO_2 inhibited the oxidation of carbon fuels however. The enhancement of fuel cell performances induced by the introduction of contaminants appears too complicated to explain. Metal cations and contaminates can also serve as mediating sites for O^{2-} exchange to promote catalysis processes. Clay materials such as kaolinite and montmorillonite contain substantial amounts of surface oxides [71], which are attractive for molten carbonate melts (hydrophilic surface functional group). These types of contaminates may enable a more intimate contact between the molten carbonate and the carbon, improving the wettability of the carbon surface. The changing of wetting conditions by coal ash contents was also reported by Allen et al. [72].

Apart from modifying carbon fuels, using carbonate melts with less surface tensions, such as smaller contacting angles, can also improve contacting conditions between solid carbon fuels and melts. Surface tensions and conductivities of molten Li-Na-K ternary systems were evaluated by Kojima et al. [73]. According to their results, surface tension parameters (γ) of Li-K binary molten systems ranged from 175 to 241 mN/m. This is lower than that of Li-Na systems (220–241 mN/m) and correlates with the contacting angle data provided by Yoshikawa et al. [61]. The data of surface tensions and equivalent conductivities can assist in the selection of proper molten mixtures. Kouchachvili et al. [74] reported an increase in peak power densities from 7.5 to 27 mW/cm^2 for a petroleum-coke-fueled MC-DCFC operating at 700°C by introducing Cs_2CO_3 into a 42.5-31.5-25 mol.% Li-Na-K ternary molten carbonate system. The concept of introducing Cs or Rb carbonates into existing molten systems was initially proposed by Kojima et al. [75] to decrease the surface tension of molten mixtures, lowering melting temperatures and increasing gas solubilities.

5.5.2 Carbon Conversion Mechanisms in Molten Carbonate

An MC-DCFC consisting of an Ni–YSZ anode and a 62–38 mol.% Li–K carbonate was constructed by Nabae et al. [9] and reached an OCV of 1.2 V at 700 and 800°C. Similar OCV values were observed by Jiang et al. [11] based on the same electrochemical system. Nabae et al. [76] suggest that the reason for these ultra-high OCV values is due to the interactions of carbon as well as CO and Li cation containing species, in which carbon and CO react with lithium oxides when operating at elevated temperatures (i.e., 700°C) through Equations 5.40 and 5.41:

$$C + O_2 + Li_2O = Li_2CO_3 \qquad (5.40)$$

$$2CO + O_2 + 2Li_2O = 2Li_2CO_3 \qquad (5.41)$$

The Gibbs free energy change of Equation 5.40 is −471.0 kJ/mol. According to this reaction, the OCV of the fuel cell can rise to 1.22 V if it dominates the anodic reaction. For the CO-related reaction shown in Equation 5.41, the OCV can rise to 1.42 V for its large free energy change of −547.0 kJ/mol. The lithium oxide in the reactions shown in these two equations can be produced by the equilibrium of carbonate decomposition found in carbonate melts, as shown in Equation 5.42:

$$Li_2CO_3 \Leftrightarrow Li_2O + CO_2 \qquad (5.42)$$

These OCV values are higher than that of carbon oxidation in molten carbonate systems and demonstrate the strong ability for these fuel cells to power external circuits. This can be explained by the thermal effects of the reactions taking place at the anode. The enthalpy change of Equation 5.42 at 700°C is +206.3 kJ/mol and the Boudouard reaction triggering Equations 5.40 and 5.41 absorbs 170.9 kJ of heat to produce two moles of CO at a temperature of 700°C, indicating that both reactions are endothermic. Therefore, the small bottom cells tested in literature need to be supplied with heat to maintain temperatures for consistent fuel cell working potentials. Overall, the reaction system absorbs heat from surrounding environments and stores it in chemical species before converting it to power. From an energy conversion point of view, this hybrid system can convert both chemical energies of carbon and heat absorbed from heating elements into electricity. Although molten carbonates, commonly considered to be ionic-liquid-containing carbonate ions $\left(CO_3^{2-}\right)$, are critical to the process, the existence of reactive oxygen ions in the medium is also important. A detailed carbon conversion mechanism was proposed by Cherepy et al. [6], which was similar to the sacrificial carbon electrode consumption mechanism in aluminum metallurgy. In this proposed mechanism, carbon oxidation is initiated by O^{2-} produced by carbonate decomposition, as shown in Equations 5.23. Free O^{2-} in the carbonate melt then adsorbs on the reactive sites of carbon surfaces for continued reactions (the adsorbed carbon atoms are noted as C_{RS}, often referring to a carbon atom at the edge, defect, step, or other surface imperfections of the carbon surface, and are more reactive than other carbon atoms). The adsorption can be expressed as Equation 5.43:

$$C_{RS} + O^{2-} = C_{RS}O^{2-} \qquad (5.43)$$

Following Equation 5.43 are two fast reactions, each with one-electron charge transfer:

$$C_{RS}O^{2-} = C_{RS}O^- + e^- \qquad (5.44)$$

$$C_{RS}O^- = C_{RS}O + e^- \qquad (5.45)$$

A second O^{2-} then adsorbs at the existing $C_{RS}O$ site, which is considered to be the rate-determining step of the whole carbon oxidation process, as shown by Equation 5.46:

$$C_{RS}O + O^{2-} = C_{RS}O_2^{2-} \qquad (5.46)$$

Charge transfer reactions similar to Equations 5.44 and 5.45 take place after the adsorption to complete the oxidization of carbon, releasing CO_2. The rest of the carbon oxidation process is shown by Equations 5.47 and 5.48:

$$C_{RS}O_2^{2-} = C_{RS}O_2^- + e^- \qquad (5.47)$$

$$C_{RS}O_2^- = CO_2(g) + e^- \qquad (5.48)$$

This mechanism was further developed by Cooper et al. [77] by adding a CO desorption process (Equation 5.49):

$$C_{RS}O = CO(g) \qquad (5.49)$$

The mechanism discussed above reveals the different reactions taking place under different overpotentials and explains the existence of so-called "low current segment" and "high current segment" in current-voltage (I-V) curves often observed in MC-DCFC testing [78–80], as shown in Figure 5.9.

At the beginning of these I-V tests, the overpotential of the fuel cells is small, and both CO_2 and CO are produced via Equations 5.48 and 5.49, leading to an increased CO_2 content in the pores of the carbon. This accumulation of CO_2 suppresses the decomposition of carbonate ions and leads to a decrease in O^{2-} concentrations. The lowered O^{2-} concentration in the reaction medium then decelerates the rate of adsorption, as shown by Equation 5.43. Therefore, the consumption of the initial absorbed species suppresses adsorption, causing gas products to block mass transport

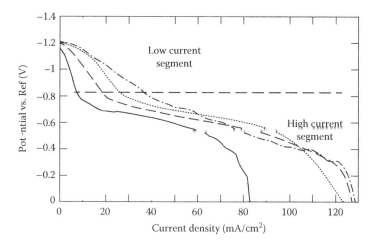

FIGURE 5.9 "Low current segment" and "high current segment" observed in a series of MC-DCFC IV curves using different carbon fuels [81]. (Reprinted with permission from Li, X. et al., Factors that determine the performance of carbon fuels in the direct carbon fuel cell, *Ind. Eng. Chem. Res.*, 47(23), 9670–9677. Copyright 2008 American Chemical Society.)

pathways, increasing impedance in the "low current region." As the working potential of the MC-DCFC decreases, overpotentials increase, causing CO accumulated at the TPB and its adjacent regions to impede the CO desorption process (Equation 5.49). As a result, the secondary adsorption process shown by Equation 5.46, considered to be a rate-determining step, is promoted by increasing $C_{RS}O$ species, causing the impedance of the fuel cell in this region to decrease. The large impedance found in the low working potential region (below 0.3 or 0.4 V) clearly shows losses caused by mass transport.

As discussed, carbon oxidation requires the existence of oxygen ions (O^{2-}). Therefore, the continuous conversion of carbon requires sufficient O^{2-} supplements. This can be achieved by oxygen decomposition at the NiO cathode of conventional MCFCs. For hybrid DCFCs with solid oxide electrolytes and carbonate melts in series however, heterogeneous transport of O^{2-} is required to provide O^{2-} to carbonate melts in the anode region. O^{2-} transport under different interfacial conditions was investigated by Xu et al. [15]. They tested three different types of anodes: a biphasic tape-casting anode made with homogeneous Ni-SDC slurry, an Ni-infiltrated anode on an SDC scaffold, and a single-phase blank SDC scaffold anode. All three anodes were immersed in a carbon-containing slurry with 62–38 mol.% Li-K molten carbonate. Peak power densities of these three types of anodes in DCFCs were completely different from those of gaseous fuels. The infiltrated anode, which was supposed to possess the best electrochemical performance because of high TPB densities [82], showed the lowest power density of 36.0 mW/cm^2 at 650°C when fueled with carbon. The blank SDC anode without an Ni–SDC interface inside, being unable to electrochemically oxidize any gaseous fuel, showed the best performance of 40.5 mW/cm^2 at the same temperature however. The peak power density of the tape-casted Ni/SDC anode was 36.7 mW/cm^2 under the same working conditions. The power densities of the three

DCFCs show that the conversion mechanisms of carbon are completely different from those of gaseous fuels. Taking H_2 conversion as an example, H_2 must diffuse into the porous electrode to reach the TPB before electrochemical reactions can take place. Therefore, high TPB densities in anodes lead to better electrochemical performances. However, it is difficult for solid carbon particles to diffuse into the pores of anodes, and instead of carbon oxidation at the TPB, carbon fuels are more prone to be oxidized by the reactive species in the carbonate melt (i.e., O^{2-}). As a result, TPB densities in the anodes are no longer dominating parameters for DCFC performances, and interfacial conditions between ionic-conducting phases and molten carbonates become key factors. The heterogeneous transport of O^{2-} under different solid–liquid interfacial conditions is shown in Figure 5.10.

For the infiltrated anode, shown in Figure 5.10b, O^{2-} transport from the SDC to the molten carbonate is blocked by Ni particles covering the surface of the SDC

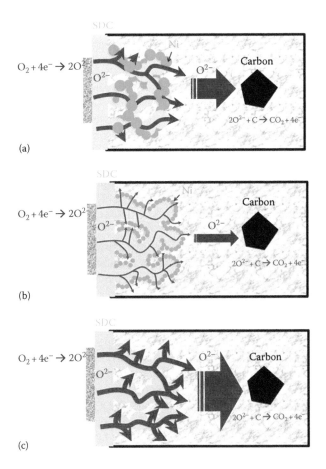

FIGURE 5.10 Oxygen ion (O^{2-}) transport under different interfacial conditions. (a) Biphasic tape casted anode, (b) infiltrated anode, and (c) blank SDC anode. (Reprinted from *Int. J. Hydrogen Energy*, 38, Xu, X. et al., Optimization of a direct carbon fuel cell for operation below 700°C, 5367–5374, Copyright 2013, with permission from Elsevier.)

framework. In the case of the blank SDC scaffold however, shown in Figure 5.10c, O^{2-} transport between the SDC and the molten carbonate is much more straightforward, allowing sufficient supplies of O^{2-} into the carbonate melt for carbon conversion, leading to the best performance among the three. For the biphasic tape-casted anode, O^{2-} transport conditions are better than the infiltrated anode but less effective than the blank SDC scaffold. These three cases demonstrate that Ni phases in solid-state anodes are not required in carbonate solid oxide hybrid systems if current collection is completed by a separate current collector. This combination of solid oxide electrolyte with molten carbonate anode was studied by Lipilin et al. [83,84], in which they inserted YSZ blind tubes into Li-Na-K ternary molten carbon melts and used noble metal meshes as anodic current collectors. In their study, they were able to obtain peak power densities of 110 mW/cm^2 at 950°C.

5.5.3 Chemical and Electrochemical Reactions in Liquid Metal

In SO-DCFCs employing liquid metals as carbon conversion electrodes, reactions can be classified into two categories: (1) oxidation of metals to oxides mentioned previously by Equation 5.24 and (2) carbon-related reduction processes. Based on current liquid metal anode studies, the electrochemical oxidation of metals still dominates interfacial regions and is the major contributor of the available current drawn from fuel cells. Therefore, the OCV of liquid metal anode DCFCs is always lower than that of direct oxidation of carbon. Due to this, investigations into metal oxidation kinetics will assist in achieving better fuel cell performances. In carbon-related reactions, apart from reducing metal oxides formed during fuel cell operations, proper carbon fuel delivery methods need to be developed to prevent the formation of isolating oxides affecting fuel cell performances (i.e., the situation of liquid Sn anodes, as discussed previously) and to increase the theoretical efficiency of liquid metal anode DCFCs.

5.5.3.1 Metal Oxidation in Liquid Metal Anodes

The oxidation of metal by O^{2-} is a heterogeneous reaction that takes place at the anode–electrolyte interface. Therefore, morphologies of the interface between the solid oxide electrolyte and the liquid metal anode play an important role in the kinetics of metal oxidation. Wang et al. [85] fabricated a liquid Sb anode DCFC based on electrolytes with different surface roughness. In their configuration, the liquid anode is in direct contact with the solid-state electrolyte, allowing the surface roughness of the electrolyte to become a parameter describing the interfacial morphology between the liquid metal anode and electrolyte. The "rough electrolyte" used was an YSZ pellet with an Ra (roughness average) value of 540 nm, while the "smooth electrolyte" used had an Ra value as small as 0.69 nm. During their fuel cell operation, carbon fuels were excluded from the anode chamber to simplify the reacting system, with the only reaction occurring at anode–electrolyte interface being the oxidation of metallic Sb. According to the inventors of the liquid Sb anode [86], metal Sb is oxidized into Sb$_2$O$_3$ through Equation 5.50:

$$2Sb + 3O^{2-} = Sb_2O_3 + 6e^-$$

(5.50)

The exchange current densities of the two fuel cells with different electrolyte roughness were calculated through the Tafel equation fitting, with the exchange current density of the liquid Sb anode using the smooth electrolyte being 1.5 mA/cm² and the rough anode–electrolyte being 2.0 mA/cm². Therefore, by increasing interfacial roughness, the exchange current densities of anodes can be increased by 33%. The peak power density of the rough electrolyte DCFC can reach 31.5 mW/cm² at 800°C, while the peak power density of the smooth electrolyte can only reach 15 mW/cm² under the same temperature.

Sb_2O_3 formed during Sb oxidation is an O^{2-} ionic conductor, according to Van Arkel et al. [87]. The ionic conductivity of Sb_2O_3 as a function of temperature is described by Equation 5.51, where $\sigma_{Sb_2O_3}$ is the ionic conductivity of Sb_2O_3 (with a unit of S/m), and T is the given temperate in Kelvin:

$$\sigma_{Sb_2O_3} = 10^{-2} \times \exp\left(-128.78 + \frac{2.9859 \times 10^5}{T} - \frac{1.7571 \times 10^8}{T^2}\right) \quad (5.51)$$

As calculated by Equation 5.51, the ionic conductivity of Sb_2O_3 at 800°C (1073 K) is 4.42 S/m. In contrast, the ionic conductivity of YSZ at 800°C (1073 K) is 2.23 S/m, as calculated by an equation provided by Shi et al. [3]. Sb_2O_3 can therefore be regarded as an ionic-conducting phase for anodes. Sb_2O_3 droplets in the anode of a bottom fuel cell can possess close contact with both YSZ electrolytes and Sb metals, forming ionic-conducting pathways. With Sb_2O_3 as the ionic-conducting phase and Sb as the electronic-conducting phase and reactant, the interface between Sb_2O_3 and metal Sb is more likely to become the electrochemical reactive sites. With the help of this newly formed reactive interface, smooth electrolyte DCFC performances can be improved. As shown in Figure 5.11a, performances increased at the beginning of the constant potential discharge and reached peak values with continued discharge in a fuel cell fabricated with a smooth electrolyte. (The case of rough electrolytes will be discussed in the latter part of this section.) Postmortem SEM photos of the interfacial region (Figure 5.11b) showed that Sb–Sb_2O_3 mixtures were found at the anode–electrolyte interfacial region (determined by Sb and O elemental distributions, as shown in Figure 5.11c and d), indicating the existence of the Sb–Sb_2O_3 interface. The formation process of the Sb–Sb_2O_3 interface as well as the electrochemical reactive sites inside the liquid Sb anode was simulated using the percolation theory in a mathematical model by Cao et al. [88].

A possibility that can occur in bottom cell configurations is that Sb_2O_3 may float up and detach from the surface of the electrolyte, because the density of metal Sb is higher than that of Sb_2O_3. However, the driven forces of Sb_2O_3 lifting are insufficient, because the buoyancy of liquid metals comes from the pressure differences between the bottom and top of the Sb_2O_3 droplet caused by the weight of the continuous liquid metal Sb surrounding it. Therefore, to lift Sb_2O_3 in a liquid Sb bath, the existence of metal Sb beneath Sb_2O_3 is necessary. Sb_2O_3 at the anode–electrolyte interface is actually attached to the electrolyte [90] as long as the electrolyte is horizontal. With the help of Sb_2O_3 connecting to the YSZ electrolyte, the electrochemical oxidation of Sb is no longer restricted to the anode–electrolyte interface. In addition, by manipulating

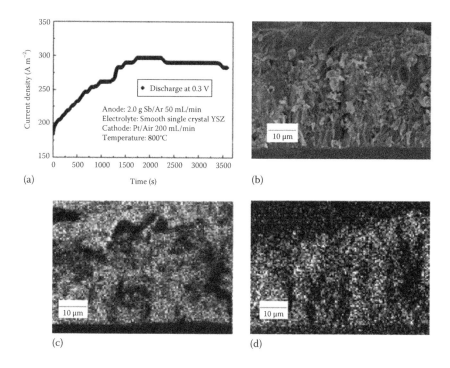

FIGURE 5.11 Performance of liquid Sb anodes and interfacial distribution of elements. (a) Current density of a fuel cell discharged at a constant potential of 0.3 V; (b) SEM photo of the anode–electrolyte interface; (c) Sb element distribution at the interfacial region presented by energy-dispersive x-ray spectroscopy, white dots as Sb element; (d) O elemental distribution at the interfacial region, white dots as O element. [89]. (Reprinted from *J. Power Sources*, 245, Wang, H., Shi, Y., and Cai, N., Polarization characteristics of liquid antimony anode with smooth single-crystal solid oxide electrolyte, 164–170, Copyright 2014, with permission from Elsevier.)

the Sb_2O_3 content at the interfacial region, the performance of the liquid Sb anode can be controlled and improved. Sb_2O_3 in liquid Sb can be electrochemically reduced (electrolyzed) into metal Sb as demonstrated by Javadekar et al. [86], demonstrating the existence of oxygen ion transport pathways inside the liquid Sb anode. Sb generation using electrolysis allows for the storage of power in the form of chemical energy in liquid metal anodes and can meet energy storage market demands.

Wetting issues can also be found in liquid Sb anodes due to the relatively large surface tensions of molten Sb (375.3 mN/m [91]). Because of this, it is hard for liquid Sb to reach the notches inside the rough surfaces of YSZ electrolytes. This wetting issue was also found in both bottom and tubular liquid Sb anode DCFCs [85,92]. This issue can be solved by Sb_2O_3 generated during fuel cell configurations. The surface tension of molten Sb_2O_3 is much smaller than that of liquid Sb metals (98 mN^{-1} at melting point [93], even smaller at elevated temperature), allowing it to wet the rough surfaces of YSZ electrolytes easier. Considering its ionic conductivity, the Sb_2O_3 between the YSZ and the metal Sb can provide better contact conditions between the electrolyte and anode. Similar metals such as Bi were also tested [94], in which the oxidation product, Bi_2O_3, is a well-known ionic conductor. The OCV of the liquid Bi anode was found to be unacceptably low however.

Other liquid metal anodes were found to be limited by isolated metal oxides deposited on the anode–electrolyte interface that blocks both charge and mass transfer [95]. Raising operating temperatures could be a possible solution to this however. A liquid Pb anode fuel cell was fabricated and tested by Jayakumar et al. [20] and it was observed that the fuel cell performance met sudden drops due to the blocking effects of PbO formation at the anode–electrolyte interface when the fuel cell operated under 700°C and 800°C. When the operation temperatures increased to 900°C however (higher than the melting point of PbO, 888°C), the sudden drop in fuel cell performances disappeared, and the liquid Pb anode fuel cell achieved a power density of 200 mW/cm². For metals with highly infusible oxides such as Sn however, this method would not work properly.

Efforts have been made to avoid the blocking effects of SnO_2 films as well as to reduce SnO_2 in liquid Sn with gaseous fuels being introduced into anode systems to reduce SnO_2. However, only SnO_2 reduction by H_2 was found to be rapid enough for continuous fuel cell operations as CO took at least 45 minutes to diffuse through the liquid Sn anode with a thickness of a few millimeters [96]. Smaller currents were also drawn from the liquid Sn anodes in an attempt to achieve a delicate balance between SnO_2 migration and generation [97]. However, the transport of SnO_2 in liquid Sn is very slow, leading to current limits set by mass transport rates resulting in low power densities.

A possible method to prevent oxide films on Sn anodes is inspired by the working processes of liquid Sb anodes. During Sb anode operations, Sb_2O_3 between the anode and electrolyte can improve the contacting conditions between the metal Sb and the YSZ electrolyte. In the case of the liquid Sn anode, liquid oxygen ion mediators, such as molten carbonate, can be used to promote O^{2-} transport. With molten carbonate occupying parts of the electrolyte surface, oxidation of Sn metals can take place at the Sn–electrolyte interface and the Sn–molten carbonate surface. The SnO_2 film formed on the Sn–molten carbonate interface is less stable than that formed on the anode–electrolyte interface because of its less "solid" foundation. Due to the fluidity of the molten carbonate, deformations of molten carbonate droplets can be easily caused by mass loaded onto its surface (i.e., newly formed SnO_2 on the surface), leading to SnO_2 detachment from the Sn–molten carbonate interface. The freed SnO_2 film fractures can then be lifted by surrounding Sn metals and leave a clear reactive interface for further Sn metal oxidation.

5.5.3.2 Carbon Conversion in Liquid Metal Anodes

Carbon conversion efficiencies in liquid anode DCFCs are among the most important parameters to be evaluated in the research and development of liquid metal anode carbon conversions. As mentioned earlier, the oxidation of metal is a major contributor to the available current drawn from fuel cells. Using liquid Sb anodes as an example, the oxidation of liquid metal Sb can be written as Equation 5.52:

$$2Sb + 1.5(O_2 + 3.76N_2) = Sb_2O_3 + 5.64N_2 \tag{5.52}$$

Carbon is required to reduce the generated Sb_2O_3, and the desired complete oxidation of carbon is described in Equation 5.53 (Note that carbon is fed to reduce Sb_2O_3 produced by Equation 5.52):

$$Sb_2O_3 + 1.5C = 2Sb + 1.5CO_2 \qquad (5.53)$$

The chemical energy used to reduce Sb_2O_3 can be calculated through the enthalpy change of Equation 5.54:

$$1.5C + 1.5O_2 = 1.5CO_2 \qquad (5.54)$$

The fuel efficiency of the liquid Sb anode can be evaluated by the CO/CO_2 equilibrium during Sb_2O_3 reduction, as shown by Equation 5.55:

$$3CO + Sb_2O_3 = 2Sb + 3CO_2 \qquad (5.55)$$

When the reaction system shown by Equation 5.55 achieves an equilibrium state at DCFC operation temperatures (i.e., 800°C), the partial pressure of CO is less than 0.01 ($\frac{P_{CO}}{P_{CO_2}} = 0.07$ at 800°C; the activities of Sn and SnO_2 are considered to be the same). Therefore, it is reasonable to assume that the complete oxidation of carbon to produce CO_2 follows Equation 5.53, and that the carbon fuels can be fully oxidized in liquid Sb anodes if the fuel cell configuration is perfectly designed. The theoretical efficiency of a liquid Sb anode DCFC oxidizing metal Sb using carbon as a reductant to produce power can be calculated based on the maximum value of work. This value of work can be extracted from the electrochemical reaction energy (Gibbs free energy change of Equation 5.52) divided by the chemical energy introduced into the reaction system (enthalpy change of Equation 5.54), as shown by Equation 5.56:

$$\eta_{Theo,Sb} = \frac{\Delta G_{Sb-O}}{\Delta H_{(C-O)_{Sb}}} \times 100\% \qquad (5.56)$$

where

$\eta_{Theo,Sb}$ denotes the theoretical efficiency of the liquid Sb anode DCFC

ΔG_{Sb-O} denotes the Gibbs free energy change of Equation 5.52

$\Delta H_{(C-O)Sb}$ denotes the enthalpy change of Equation 5.54

Fuel efficiencies must be considered if more reducible metals, such as Sn, are applied as the liquid metal anode. The gaseous-phase equilibrium between CO and CO_2 is evaluated in Equation 5.57:

$$2CO + SnO_2 = Sn + 2CO_2 \qquad (5.57)$$

To fully reduce 1.0 mol of SnO_2, more than 1.0 mol of carbon must be fed to the reaction system because a portion of the carbon is partially oxidized to CO and cannot be further oxidized. The amount of carbon required to reduce 1.0 mol of SnO_2 is expressed by Equation 5.58:

$$\left(\frac{2+2\exp\left(-\dfrac{\Delta G_{Sn-CO}}{2RT}\right)}{1+2\exp\left(-\dfrac{\Delta G_{Sn-CO}}{2RT}\right)}\right)C+SnO_2=Sn+\left(\frac{2}{1+2\exp\left(-\dfrac{\Delta G_{Sn-CO}}{2RT}\right)}\right)CO$$

$$+\left(\frac{2\exp\left(-\dfrac{\Delta G_{Sn-CO}}{2RT}\right)}{1+2\exp\left(-\dfrac{\Delta G_{Sn-CO}}{2RT}\right)}\right)CO_2 \quad (5.58)$$

where ΔG_{Sn-CO} is the Gibbs free energy change of Equation 5.57. van't Hoff's law is applied to evaluate the equilibrium atmospheres of Equation 5.57, and the activities of Sn and SnO_2 can be expressed by Equation 5.59:

$$2\ln\left(\frac{P_{CO_2}\big/P_\theta}{P_{CO}\big/P_\theta}\right)=-\frac{\Delta G_{Sn-CO}}{RT} \quad (5.59)$$

The overall reaction of metal Sn oxidation by air can be expressed by Equation 5.60:

$$Sn+(O_2+3.76N_2)=SnO_2+3.76N_2 \quad (5.60)$$

The chemical energy required to reduce SnO_2 produced by Equation 5.60 is evaluated using the enthalpy change of Equation 5.61, and the molar amount of carbon is determined by Equation 5.60:

$$\left(\frac{2+2\exp\left(-\dfrac{\Delta G_{Sn-CO}}{2RT}\right)}{1+2\exp\left(-\dfrac{\Delta G_{Sn-CO}}{2RT}\right)}\right)C+\left(\frac{1+\exp\left(-\dfrac{\Delta G_{Sn-CO}}{2RT}\right)}{1+2\exp\left(-\dfrac{\Delta G_{Sn-CO}}{2RT}\right)}\right)O_2$$

$$=\left(\frac{2+2\exp\left(-\dfrac{\Delta G_{Sn-CO}}{2RT}\right)}{1+2\exp\left(-\dfrac{\Delta G_{Sn-CO}}{2RT}\right)}\right)CO_2 \quad (5.61)$$

The theoretical efficiency of a liquid Sn anode DCFC can be expressed in the same manner as the liquid Sb anode discussed earlier, shown by Equation 5.62, where $\eta_{Theo,Sn}$ denotes the theoretical efficiency of the liquid Sb anode DCFC, ΔG_{Sn-O} denotes the Gibbs free energy change of Equation 5.60, and $\Delta H_{(C-O)Sn}$ denotes the enthalpy change of Equation 5.61.

$$\eta_{Theo,Sn} = \frac{\Delta G_{Sn-O}}{\Delta H_{(C-O)_{Sn}}} \times 100\% \qquad (5.62)$$

The theoretical efficiencies of the intensely studied liquid Sb and liquid Sn anodes are calculated based on Equations 5.56 and 5.62, with the results shown in Figure 5.12. Gaseous equilibrium data calculated based on Equations 5.55 and 5.57 are also presented in Figure 5.12 as an important parameter during fuel cell efficiency evaluations. Figure 5.12 shows that in regard to OCV values, the theoretical efficiencies of liquid Sn anodes are much higher than that of liquid Sb anodes. These higher OCV values also mean more CO content in the flue gas however.

Carbon materials [98,99] as well as heavy carbonaceous fuels [100] can be used as feeding fuels for liquid metal anodes. Jayakumar et al. operated a liquid Sb anode DCFC using an ScSZ electrolyte at 700°C with biomass and charcoal as feeding fuels. During a constant working potential (0.5 V) discharge test running for 200 hours, the fuel cell exhibited stable power densities of 230–290 mW/cm². The OCV value was found to be dominated by Sb oxidation because metal Sb can occupy the anode–electrolyte and the carbon can alternatively serve as the reductant of Sb_2O_3. A short-lived high OCV of over 1.0 V was observed by Cao et al. [101] in a carbon–Sb anode operated

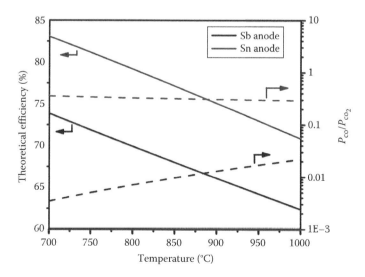

FIGURE 5.12 Theoretical efficiencies of liquid Sb and liquid Sn anode DCFCs (oxidizing metal and fed with carbon) and equilibrium gas contents over metal oxide reductions as a function of temperature.

at 800°C. The short-lived high OCV value suggests the existence of a highly nonequilibrium state inside the liquid Sb anode. This high OCV value suggests that carbon or carbon-related species play a role in the electrochemical reactions when presented to the anode–electrolyte interface. Xu et al. [23] studied the reactivity of several carbon fuels toward SnO_2 reduction. When the most reactive activated carbon was mixed with Sn, the OCV of the liquid Sn anode DCFC achieved a high and steady value of 1.0 V with a peak power density of 133.0 mW/cm^2 under 800°C. This activated-carbon-loaded liquid Sn anode DCFC was then discharged at a constant potential of 0.5 V at 850°C and produced a steady power density of 92.5–107.5 mW/cm^2 over 10 hours of discharging. Postmortem SEM and energy dispersive x-ray spectroscopy analysis confirmed that the SnO_2 was reduced by the highly reactive activated carbon and no SnO_2 blocking films were found deposited on the electrochemical reactive surface. The well mixing of fuels and liquid metals was achieved in a tubular liquid Sb anode DCFC designed by Cao et al. [92,102]. In their experiment, de-ash coal was carried beneath the liquid Sb surface by an Ar carrier gas, and coal suspension in the liquid Sb bath was achieved. With the fuel being in direct contact with the anode–electrolyte interface, a high OCV of 0.83 V was obtained. Mass transport of fuels in tubular liquid Sb anodes was found to be different when compared to bottom-type fuel transports as mentioned earlier. For vertical anode–electrolyte interfaces, lifting of Sb_2O_3 formed at the interface by Sb is more significant than that in bottom-type fuel cells [103]. Fluidization of liquid Sb baths can improve fuel cell performances by enhancing mass transport inside anodes. The usage of large-dimensional fluidized anodes makes it easier to arrange fueling equipments, enabling the continuous introduction of carbon fuels into the anode. Power outputs of the system can be easily scaled up by inserting more tubular fuel cells into the liquid Sb bath.

A tubular liquid Sn anode fuel cell was developed by Tao et al. [99,104] (known also as CellTech Power). One feature of this liquid Sn anode fuel cell is a porous ceramic separator to maintain a liquid Sn thin layer attached to the YSZ electrolyte with a thickness of a few hundred micrometers. With a pore diameter of 100 μm, liquid Sn is trapped between the electrolyte and the separator. The average porosity of the separator is more than 60% [104] and allows fuel molecules to diffuse into the anode. The cathode of this DCFC is the inner tube that the gas-tight YSZ electrolyte is coated on. During fuel cell operations, liquid tin anodes located between the electrolyte and the porous separator are oxidized by oxygen ions transported from the electrolyte, while at the same time, generated SnO_2 is reduced at the anode–electrolyte interface by reducing species passing through the thin layers of Sn. A detailed structure of this liquid tin anode tubular fuel cell is presented in Figure 5.13.

One impressive characteristic of the system is its flexibility for various fuels. Small molecules (H_2, natural gas), heavy carbonaceous fuels (JP-8 kerosene, biodiesel), and solid-state carbon materials can all be consumed to produce power. Liquid Sn anode fuel cells have also shown satisfying sulfur tolerances in which no noticeable performance degradation was observed in a 200 hours sulfur tolerance test where JP-8 (Jet Propellant 8, a jet fuel widely used by the U.S. military) with a sulfur content of 1400 ppm was fed into the anode chamber as the fuel, without desulfuration or other fuel pretreatment processes. Sulfur, according to Tao et al. [99], is a fuel rather than a poison in liquid Sn anodes. For Gen3.1-type fuel cells, 3 W of

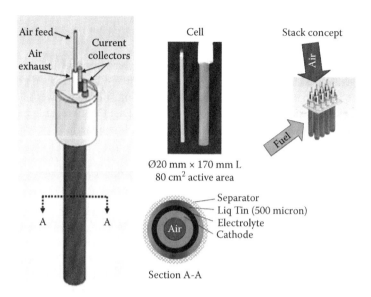

FIGURE 5.13 Liquid tin anode fuel cell design, from single cell to stack [105]. (From Carlson E.J. et al., *Assessment of a Novel Direct Coal Conversion—Fuel Cell Technology for Electric Utility Markets*, Electric Power Research Institute, Palo Alto, CA, Copyright 2006 Electric Power Research Institute.)

power can be extracted from a 10 cm long, 1.0 cm round, single tubular unit. Fuel cell power densities of this liquid Sn anode fuel cell can increase from 40 to 150 mW/cm^2 operating under 1000°C when fueled with JP-8.

The discussions earlier mainly address reactive metal anodes. For inert liquid metal anodes, such as liquid Ag anodes, carbon is oxidized by the oxygen dissolved in liquid Ag rather than by silver oxides. This is because Ag_2O is thermodynamically unstable at the melting temperature of silver. Mechanisms of carbon conversion in liquid silver were proposed by Gopalan et al. [106], in which oxygen ions in the YSZ lattice can dissolve in liquid Ag (noted as O(Ag)), as shown by Equation 5.63:

$$O^{2-}(YSZ) = O(Ag) + 2e^- \qquad (5.63)$$

Both the partial and complete oxidation of metal silver can take place at elevated temperatures, as shown by Equations 5.64 and 5.65:

$$C + O(Ag) = CO \qquad (5.64)$$

$$C + 2O(Ag) = CO_2 \qquad (5.65)$$

As mentioned in previous sections, the slow migration of dissolved oxygen can hinder the whole reaction process. However, it is still worthwhile to improve a method to accelerate oxygen transport in liquid silver to help achieve "direct" carbon oxidation.

5.6   DEVELOPMENT OF CARBON-BASED FUEL CELL STACKS AND SYSTEMS

Development of carbon-based fuel cell stacks is necessary for practical applications of power generation using solid carbon as fuels. Unlike other advanced coal-based energy pipelines, such as IGCC or supercritical PC, most of the power is extracted from the fuel cell instead of the combustion process. DCFCs are regarded as an efficient and clean energy technology to convert solid carbon materials.

5.6.1   DIRECT CARBON FUEL CELL STACKS AND ASSOCIATED SYSTEMS

As discussed earlier, although many research works have demonstrated carbon conversion in DCFCs, few fuel cell configurations have shown viability in terms of scaling and continuous operation. Among various DCFC fuel cell configurations, the Lawrence Livermore National Laboratory (LLNL) has demonstrated a promising planer fuel cell configuration [107] with an arrangement of 150 cm^2-planer MCFCs. Their fuel cell configuration can generate as much as 30 W DC power when operating under 750°C. When employing molten carbonates as the electrolyte, the contamination of the electrolyte by ash (usually calcium sulfate and calcium aluminum silicates are the dominating soluble minerals) in coal requires consideration, and the electrolyte should be replaced to maintain fuel cell performances when ash contents exceed 10% in the carbonate melt. Therefore, LLNL's MC-DCFC system had to be fed on de-ash coal with total ash contents ranging from 0.1% to 0.2%. In case of 0.18% ash content, the electrolyte of the fuel cell had to be replaced every 1.5 years. According to an Electric Power Research Institute report [108], the fuel expense of raw coal (Ohio coal) is 1.42 (2006$) GJ^{-1} (original value, 1.5 (2006 U.S. dollar) $MBtu^{-1}$), whereas the fuel expense of de-ash coal is 3.98 (2006$) GJ^{-1}. This dramatic increase in fuel prices leads to significant increases in the cost of electricity, in which the cost of de-ash coal took up 36.8% of the total power generation costs (which was 59 (2006$) MWh^{-1} based on a calculation without considering CO_2 sequestration).

Planar fuel cell configurations possess stacking potentials, in which voltage can be built up by connecting bipolar fuel cells in series or in parallel to enlarge current output. However, planar fuel cell configurations possess several issues such as difficult sealing, difficult continuous feeding of carbon fuel into the thin anode cavities, impossible replacement of molten carbonate electrolytes during fuel cell operations due to the soaking of carbonate melts into the electrolyte-supporting matrix, as well as low CO_2 sequestration ratios.

Other DCFC systems use tubular fuel cell configurations with solid oxide electrolytes. An SO-DCFC based on carbon fluidization was demonstrated by Li et al. [53]. Their fuel cell performance was limited by the ultrathick (1.3 mm) YSZ tube employed as the SOFC electrolyte. The fuel cell reached a peak power density of 22 mW/cm^2 at 900°C using the same He fluidized bed when a 0.2 mm thick YSZ disk was used as the electrolyte. The peak power density of the DCFC further improved to 43 mW/cm^2 when the fluidizing gas was changed to CO_2 with another thin YSZ-disk-supported fuel cell operating at 900°C [52]. Peak power densities of 200 mW/cm^2 can also be achieved with anode supported fuel cells [109].

These reported power densities demonstrate the possibility of scaled-up DCFCs with fluidized anodes. Larger power outputs can be achieved by inserting several tubular SOFCs into a fluidized bed reactor without sealing issues, and the voltage can also be increased by connecting the fuel cells in the same reactor in series mode. As discussed earlier, the stacking of fuel cells requires a larger fluidized bed, and using a large-diameter reactor benefits both continuous fuel feeding and ash handling, both of which are common technologies in boilers. Fluidization of DCFC anodes can also provide other benefits, such as enhanced heat transfer in the carbon bed through fluid convection as the temperature difference between the upper and lower surfaces of an 8 mm thick fixed carbon bed can be as large as $100°C$ [110]. Fluidization can also mitigate temperature differences between the carbon-loaded section and the freeboard section in a single, tubular fuel cell. Fluidizing carbon beds in anode chambers can be achieved by the relatively simple circulation of anode flue gases. The development of SO-DCFC stacks is currently impeded by the limited selection of commercialized single cells with electrolyte supported tubes originally designed for oxygen sensing as well as tubular fuel cells supported by inner anodes (most tubular fuel cells available use inner tube anodes to seal reducing atmospheres in the tube). Due to this limitation, the advantages of fluidized anodes are not well demonstrated. There are also issues regarding the fluidization of carbon beds. Although the performance degradation of fuel cells is attributed to sulfur poisoning of Ni anodes [53], abrasion of anodes is more likely to be the main reason for power density shrinkages and requires careful fluidization engineering (i.e., bubbling bed) to prolong fuel cell stack lifetimes.

Other scaled-up DCFC stacks adopt YSZ tubes with molten media. For example, the Stanford Research Institute (SRI) combined SOFC technologies with Li-Na-K ternary molten carbonates [83]. Their latest design consists of an inner tube cathode supporter lending strength to the fuel cell and a thin layer of YSZ serving as the electrolyte, without the solid-state anode commonly found in SOFCs. This half-cell inserted into a carbonate melt can achieve a power density of 280 mW/cm^2 when operating under $700°C$ [108]. Although this system also uses molten carbonate as the carbon conversion media, the gas-tight YSZ makes it unnecessary to supply CO_2 to the cathode. Ash handling in this system is also much easier compared to the LLNL technology because circulation of the carbonate melt in the tubular reactor is possible in this configuration. Contaminated molten carbonate is released from bottom of the fuel cell reactor because the ash content is heavier than molten carbonate (around 3.3 t/m^3 vs. around 2.4 t/m^3), which then undergoes a de-ashing process before being circulated back into the reactor. The circulation of molten carbonate also provides opportunities for temperature control during fuel cell system operations. Excess amounts of heat released by the electrochemical reactions can be recovered by a heat exchanger, in which the molten carbonate flow is cooled down to $450°C$, while the working fluid of the bottom cycle is heated up. Agitation anode fluidization of the molten carbonate is also allowed to enhance mass transport of fuels. Due to the ionic conductivity of molten carbonates, it is hard for several fuel cells sharing the same molten carbonate bath to build up voltage. Therefore, the fuel cells in a single reactor in this SRI technology are connected in parallel to enlarge power outputs, and several reactors have to be connected in series to provide higher working stack voltages.

Liquid-tin-based DCFC stacks were proposed by CellTech power LLC [111,112] as a combination of SOFC technology and liquid tin anodes. For the sub-kW-scale liquid tin anode DCFC stack system, coal is first gasified in the anode chamber of the DCFC system. The gaseous products are then permeated through a porous ceramic separator to engage in the power generation process. The anode flue gases containing CO_2 are then circulated back into the carbon bed as a gasification agent. For larger systems, increasing the number of assembled tubes will require simplifications to the single cell. Therefore, the ceramic separator of every cell, as mentioned previously, is no longer needed. Instead, cathode-supported tubes with thin YSZ coatings on the outside are directly immersed in a 1000°C liquid tin bath. The metal tin is electrochemically oxidized when power is extracted from the fuel cell system. The oxygen content in the metal is controlled to under 0.1 wt.% to avoid the formation of dense SnO_2 films. The liquid tin metal in this configuration circulates in the same manner as in the SRI technology in which the oxygen-saturated liquid tin results in a standalone tin reduction reactor in which carbon fuels are fed as a reductant. Because the reduction of SnO_2 is an endothermic reaction, the temperature of the liquid tin flow is lowered to 973°C. The ash content of coal is removed by procedures similar to those used in modern floating-glass industries where coal ash/slag is floated over molten tin (around 3.3 t/m^3 vs. around 6.6 t/m^3) and is skimmed off as most oxides and salts have little or no solubility in molten tin. Therefore, contaminates of coal are isolated from the anode chamber and elements like sulfur are enriched in the tin reduction reactor. The reduced tin flow is then further cooled down to 961.7°C to counterbalance the heat released from tin oxidation and returned into the anode chamber. Similar circulation strategies of hot and dense fluid circulation were also proposed by Jayakumar et al. [21] in a liquid-Sb-based system. These strategies appear more persuasive on liquid Sn anodes because liquid Sn flows can be cooled down to as low as 250°C. The electronic conductivity of liquid tin can also make it impossible to build up voltages with several tubular fuel cells in the same liquid tin bath. The main challenge of this technology is the immense flow rate of liquid tin required to maintain a low oxygen content in the liquid metal. For example, in a liquid tin anode fuel cell stack with each cell operating at a typical working potential of 0.654 V, the tin flow rate has to be maintained at around 0.13 kg/s to maintain continuous power productions of 1.0 kW.

The efficiencies of DCFC systems are quite impressive, even for the most inefficient liquid tin anode DCFC technology pipelines. Overall efficiencies of DCFC systems can be as high as 61% based on the high heat values of the given coal, without considering carbon capture (57% when considered) [108]. In a DCFC stack, the heat of the cathode flue gas is usually used to preheat the fresh air that will be injected into the cathode, while the heat of the anode, usually in form of hot fluid or pyrolytic products of coal, is recovered by a bottom cycle. The bottom cycle, which is chosen to be a Rankine cycle in several reports [108,113], is often less optimized however. Normally, a small-scaled turbine operates on fluids with low temperature and pressure, i.e., a 10 MW scale industrial steam turbine requires steam at 480°C–505°C, whereas the temperature of the anode exhaust or circulated molten fluid in DCFC systems is often around 700°C or higher (sometimes as high as 1000°C). As a result, large energy losses take place during the heat exchange process as the temperature difference between the heat source and the working fluid is too large. For steam turbines working at higher temperatures, i.e., 565°C–585°C, power outputs (150–250 MW) are too large for DCFC hybrid energy systems. Therefore, well-selected thermal engines will improve DCFC system efficiencies, as the operations of a bottom cycle often take up 5%–15% of a hybrid system's power output. Other factors, such as ash content handling, potential of scaling up, and flexibility of fuel, also play important roles in DCFC system designs.

Apart from bottom cycles and other BOP equipment, the fabrication of DCFC stacks is a very important aspect. Even for a small-scale 1.0 MW_{el} system, large quantities of tubes, single cells, and reactors are required. To meet the special requirements of carbon conversion anodes and supplements of fuel, cathode-supported tubular SOFC half-cells must be fabricated. According to an SRI estimation, large-scale tubular half-cells, possessing an active area of 9423 cm^2 and supplying a power output of around 2.7 kW, will cost $1000 (2006$) per single tube [108]. If the costs of current collection, enclosure, and stacking are taken into consideration as well (often taking up one-third of the stack cost [114]), the cost of a DCFC stack will greatly exceed the goals set by DOE of $448 (2006$) per kW^{-1} [115] (original data were $400 (2002$) per kW^{-1}, an inflation rate of 12.1% between 2002 and 2006 U.S. dollar has been taken into consideration). Due to the early stages of development and fabrication for DCFC systems, estimations of capital cost for DCFC systems are highly plausible. However, there are a few reasons for higher prices to be "allowed" for DCFC systems. These reasons include less fuel demands due to higher efficiencies, less labor requirements due to simpler systems, as well as less fuel processing fees and CO_2 capture benefits.

5.6.2 INTEGRATED GASIFICATION FUEL CELL SYSTEMS

Preliminary discussions and cost comparisons of IGFC systems with other advanced coal-fired technologies have been presented in previous sections, while more detailed information, including system configurations, operation parameters, and thermodynamic efficiencies of gasification-driven, carbon-fueled fuel cell hybrid energy systems, is presented in this section.

MCFCs are a more mature high-temperature fuel cell technology than SOFCs. MCFC power plants with outputs of hundreds of kilowatts to megawatts have been built and operated around the world. Therefore, hybrid systems combining MCFC systems, coal gasifiers, and other thermal engines are more likely to succeed. An MCFC–GT hybrid system fed on syngas was proposed by Greppi et al. [116] with a 125 kW MCFC stack operating at 570°C–690°C. Efficiencies based on the higher heating value (HHV) of this hybrid system increased from 33% to 43% through the optimized selection of the micro gas turbine engine. This engine, designed for 30 kW power outputs working on a 15 kW load, was more efficient than an engine working at 37 kW with a full capacity of 100 kW. Syngas-fueled systems, although optimized, are still less efficient than gaseous hydrocarbon-fueled systems. This is a result of the high H_2O content in mass circulation and the low extent of coupling between endothermic and exothermic processes. Shifting reactors for water–gas shifting reactions in the system require superheated steam at 350 kPa and 300°C. The stream is heated from 15°C liquid-phase water.

The evaporation of this mass flow demands a heating duty of 103 kW, in which the heating of the liquid from 15°C to 139°C takes up 11% and the sensible heating of the vapor from 139°C to 300°C takes up 17%. The remaining heat (72%) is used for water evaporation to overcome the latent heat of the phase transfer at 139°C. One of the main sources of the heat for this evaporation process is heat recovered from the anode flue gases at a temperature of 630°C. Although the temperatures of this stream appear sufficient, due to the high water content in the anode flue gases (around 55 mol.%), most of the heat can only be released in the form of latent heat; 169 kW of heat can be recovered if the anode flue gas flow can be fully condensed, but only 27% of the heat is released in the form of sensible heat when flue gases are cooled down from 630°C to 120°C. The rest of the energy is released in the form of condensing heat during vapor condensation between 120°C and 83°C. Therefore, the mismatch between condensation (83°C–120°C) and evaporation (139°C) temperatures of the two different streams causes direct heat recovery from high-temperature anode flue gases to be less efficient. The extra heat needed and the wasted heat inevitably lower system efficiencies.

SOFC-based systems are another technical pipeline to fulfill the demand for distributed energy supplements. Bang-Møller et al. [117] proposed a hybrid power system composed of a pressurized SOFC stack and a micro gas turbine. Wet wood chips were utilized as the feedstock of a low-tar biomass gasifier. In the original system configuration, wood chips were first converted to raw syngases by a gasifier. The raw gases were then cooled down and run through bag filters to remove particulates, condense excess water, and reduce tar. Sulfur removal was not considered in their system, as the S content was insignificant in the product gas. The purified syngas was then blown into the anode chamber by a fuel compressor and preheated by anode flue gases, while air was pressurized and preheated at the cathode side at the same time. Their fuel cell stack consisted of 75 pieces of planar SOFCs, with 81 cm^2 of reactive area each, operating at around 800°C, at an elevated pressure of 0.25 MPa. Residue fuel and air leaving the SOFC stack were mixed and burned to drive a gas turbine downstream to acquire additional power. This fuel cell stack generated 227 kW of power after DC–AC conversions, and 48 kW of power was extracted from the micro gas turbine. The total net efficiency of this hybrid system is 48.1% with respect to the higher heat value of the wood chip feedstock (13.3 MJ/kg). An energy analysis was carried out to evaluate the loss of useful energy in the system, with 27% of the total energy loss being attributed to the gasifying reactor. Another major contributor to energy loss is the various heat exchangers, with 28% of total energy lost during heat exchanges. Therefore, heat exchanger networks need to be optimized as well. In order to reduce energy loss, hot raw syngases were employed to heat inlet fuel gases before it is preheated by anode flue gases. The temperature of the purified syngas before entering the anodic heat exchanger was increased from 173°C to 498°C, and the fuel cell working pressure was increased from 0.25 to 0.275 MPa. With less heat extracted from the anode flue gases, flue gas temperatures entering the burner increased from 410°C to 675°C, leading to a higher turbine inlet temperature of 790°C (original 706°C). The flue gas leaving the hybrid system was at 272°C and was used for feedstock drying. The overall efficiency of this optimized system increased to 50.8% with power produced by SOFC remaining at 227 kW, while work done by micro gas turbine increased to 64 kW. This hybrid system demonstrates potential for fuel gas purification, efficiency of micro gas turbines, and the coupling effects of SOFC stacks and rotatory machinery. The study earlier shows that the utilization of delicate and efficient power systems such as SOFCs and micro turbines requires careful thermal designs and proper fuel pretreatments, such as advanced gasifiers.

A large centralized IGFC power plant consisting of coal gasification, and cleanup processes with an SOFC system as the top cycle and a subcritical steam Rankine cycle as the bottom cycle, was analyzed by NETL as mentioned in the previous section. In this system, purified, high-pressure (2.32 MPa) syngases are discharged through a syngas expander before entering the SOFC island. The huge power output is modularized to eight SOFC blocks with each block consisting of 42 SOFC modules. Each SOFC module is then broken down to 64 SOFC stacks containing 96 pieces of anode-supported planar SOFCs with a reactive area of 550 cm^2 each [57,118,119]. Each of the eight blocks housed individual SOFC Balance of Plants (BoP) including a blower for the cathode and anode chambers and a sand heat exchanger for reclaiming heat from flue gases. The anode off-gases from the eight blocks are collected and burned in an oxy-combustor with oxygen streams supplied by an air separation unit (ASU) to generate heat for the bottom cycle. The bottom cycle is a subcritical reheated Rankine cycle, in which main steam enters a high-pressure turbine at 12.4 MPa/559°C–562°C and reheated steam enters an intermediate pressure turbine at 3.1 MPa/558°C–561°C. The CO_2 produced during power generation is pressurized, liquefied, and purified to enhanced oil recovery levels. This analysis predicts that high efficiencies can be achieved by using richer CH_4 fuel streams for the SOFC island. Better gasifiers are also needed by the centralized energy system. The baseline case of this IGFC system uses a CoP E-Gas™ two-stage, water–coal slurry fed gasifier. The cold gas efficiency of this conventional gasifier is 81%, based on the HHV of the coal fed. The gasifier provides a syngas stream containing 5.8 mol.% CH_4 (37.7 mol.% of CO, 20.4 mol.% of CO_2, 35.2 mol.% of H_2, 0.1 mol.% of H_2O, N_2 take up the rest) and leads to a net electric efficiency (based on the HHV of Illinois No. 6 coal) of 39.5%. A total power of 729.9 MW can be generated by this IGFC system with the fuel cell generating 76%, the steam turbine generating 19%, and the syngas expander generating 5%. Parasitic loads of the baseline plant, being as high as 179.8 MW, include CO_2 sequestration (consumes 42% of auxiliary power) and air separation (31%) processes as the main in-plant power consumers. The net system efficiency can improve from 39.5% to 48.9% if CH_4 content is increased in the syngas. An Exxon [120]-derived K-based catalytic gasifier was considered to produce syngas CH_4 at lower temperatures and lower oxygen consumption rates. The catalytic gasifier provided a syngas flow containing CH_4 contents as high as 31.6 mol.% (diluted to 10.4 mol. % at anode inlet by anode flue gas circulation to prevent carbon deposition). A total power of 668.9 MW was generated by this CH_4-rich fed plant. Fuel cell performances under similar working conditions were promoted as a result of the enhanced cooling due to methane in situ reforming. The SOFC subsystem took up 88% of the total power output, whereas the parasitic load of this optimized plant shrank to 100.8 MW because the power consumed by the ASU decreased from 41.9 MW in the baseline case to 15.8 MW in the present case (characterized by the power consumption rate of the main air compressor of the ASU) as a consequence of less oxygen being needed.

In the proposed system, the degradation rates of the fuel cells play an important role in determining the cost of electricity (COE). For a fuel cell stack working at a given degradation rate, a realistic strategy of extracting constant power is to operate the stack at higher potentials than the designed working conditions from the beginning. The limit to lowering operation potentials is the oxidation limit of Ni-based anodes. This limit requires a working potential of above 0.7 V and a lower limit of 0.75 V. These values are always chosen for fuel cell operations in order to provide a marginal space of 0.05 V. Therefore, part of the stacks needs to be replaced to maintain system performances. Extra amounts of reactive areas are enclosed into the stacks to prolong its lifetime because the effects of extending lifetimes become more significant as degradation rates decrease. In current state-of-the-art SOFC technologies, degradation rates of 1.5% per 1000 hours can be achieved with lifetimes of fuel cell stacks reaching 1 year (8000 hours). If the degradation rate can be minimized to 0.2% per 1000 hours with a 10% additional reactive area installed, the lifetime of the stacks can be prolonged to around 7.3 years [121]. The difference in stack lifetimes directly determines the number of stacks to be replaced annually, affecting the COE in the form of operation and maintenance costs. The COE can be decreased to $ 46.4 (2011$) MWh^{-1} if degradation rates are lowered from 1.5% per 1000 hours (baseline case) to 0.2% per 1000 hours. This is a dramatic improvement considering that the COE of the baseline case is as high as $170.2 (2011$) MWh^{-1}.

Pressurization of fuel cell systems is not financially attractive. SOFC islands working at ambient pressures and elevated pressures (1.96 MPa) were evaluated and compared by using a catalytic gasifier as the fuel supplier mentioned earlier. Although the pressurization of SOFC subsystems can increase net electric efficiencies from 53.5% to 58.4% based on the HHV of Illinois No. 6 coal, the COE of the pressurized system is more expensive than that of the ambient pressure system by $5.1 (2011$) MWh^{-1}, negating the benefits of the higher efficiency in pressurized systems. Ambient SOFC modules cost $382 (2011$) kW^{-1}, including the cost of the SOFC stack (225 (2011$) kW^{-1}), enclosure, inverter (equipment cost and power loss during DC–AC conversion both considered), and so on. In comparison, pressurized vessels cost as high as 240 (2011$) kW^{-1}, and the total cost of pressurized SOFC modules can reach $592 (2011$) kW^{-1}.

5.7 CHAPTER SUMMARY

To meet the conflicting requirements of energy demands and environmental protection, advanced carbon conversion technologies as well as novel power grid configurations are in great need. Carbon-based fuel cells or its hybrid systems can work as complementary technology pipelines to fulfill distributed energy demands of customers and reduce or even eliminate pollutant emissions. Clean and efficient DCFC power or CHP (cogeneration of heat and power) systems can provide electricity and heat to blocks, houses, remote villages, and offshore islands. There are several potential methods for carbon-fueled fuel cell systems to work as distributed power generators.

Biomasses are relatively abundant and possess medium HHVs (15–20 MJ/kg), large moisture contents, lower ash contents, and simpler elemental compositions than coal. Therefore, biomass fuels are ideal for distributed power systems (cost of biomass collection will be offset by low fuel costs for centralized systems) and can either be digested by advanced gasifiers [122] to produce purified fuel gases for downstream fuel cell systems or be fed into the molten media of DCFCs as mentioned in previous sections. Distributed-carbon-based power generation can also be achieved with the help of modern logistics networks. Commercialized carbon fuels can be provided through e-commerce platforms, in which de-ash coal, carbon black, and purified carbon–water slurry can be ordered through the Internet and transported to customers within hours to days. De-ash carbon fuels allow for easier ash handling in both gasifiers and fuel cells, as it is less technically challenging for fuel cell systems if contaminants in carbon fuels are removed before being fed to the system.

A bolder method for distributed power generation is to connect several power systems in a neighborhood together to form a micro power grid. System stabilities would be improved by the multiple power sources and energy storages in the microgrid, while system expenses related to connecting power lines to main power girds as well as losses incurred from long-distance power transport can be minimized. This method can be realized by the batching of liquid Sb anode DCFC systems as mentioned in previous sections. Liquid Sb metal itself is a good carbon-converting electrode, and the chemical energy in metal Sb can also be oxidized as a power source. Therefore, liquid Sb anode DCFCs can maintain power production by consuming itself when there is a scarcity of carbon fuels. This can be regarded as the "battery mode" of liquid Sb anode DCFC operations [95]. The discharge of liquid Sb is reversible because the Sb_2O_3 formed during oxidation can be reduced by introducing carbon fuels. Another advantage of liquid Sb electrodes is that Sb_2O_3, a common phase in the metal bath, can be electrolyzed [86], allowing excessive amounts of generated power to be stored in the form of liquid Sb metals. Sb metal in this distributed energy system can be regarded as both a carbon converter and an energy storage media. Therefore, a microgrid consisting of several liquid Sb DCFCs can produce/store energy based on the demands of customers, while at the same time be controlled and managed by the latest information technologies or even artificial intelligence.

5.7.1 Technical Challenges of DCFCs

Apart from difficulties in carbon conversion as discussed earlier, there are additional technical challenges associated with the utilization of carbon as a fuel, including current collection, stability of molten media, and fuel cell performances under various coal contaminants.

5.7.1.1 Current Collection of DCFCs

Current collection of high-temperature fuel cells is difficult but critical in guaranteeing fuel cell performances, especially in DCFCs employing molten media as carbon conversion electrodes. For sacrificial rigid carbon rods inserted into molten media serving as both the fuel and current collector [4], there is a counterbalance between electronic conductivity and fuel reactivity. Normally, a highly regulated lattice (i.e., graphite) is preferred as it has higher electronic conductivities. But in terms of high reactivity, carbon with more microscopic defects is preferred. The same issue also occurs in MCFCs using carbon–carbonate slurries as fuel, in which piles of carbon particles serve as current collectors. Current collection in this configuration is achieved by the closely packed carbon particles [123], but the contacting resistance between the particles can be very large. Although perfectly packed bed anodic current collection was well simulated [124], model deviations from real DCFC working conditions should be considered. As carbon is consumed during DCFC operations, the radius of these carbon particles shrinks, leading to looser carbon particle packing conditions, in which smaller carbon particles in the anode will lose contact with each other, resulting in current collection failure.

Although higher electronic conductivities can be achieved by liquid metal anodes, current collection is also an issue in which the metals used for current collection are often soluble in the liquid metal anode. For example, a solution of Ni in liquid Sn at 800°C can be as high as 20 atm.%, prohibiting the use of most stainless steel materials as current collectors. Metallic elements in ferrite alloys, such as Fe, Cr, and Mn, are generally more reducing than liquid metal anode materials, and as a result, they are more readily oxidized by oxygen ions than anode metals during fuel cell operations. Thus, it is very difficult to explore the oxidation kinetics of metal materials. Rhenium (Re) is the only stable current collection material commonly used for liquid metal anodes. New current collecting methods have to be developed if larger-scale stacks, consisting of several independent reactors, are to be built.

Research on DCFCs has often focused on the mechanisms of carbon conversion or the development of carbon-converting electrode materials. Few novel SOFC structural designs have ever been applied in the field of SOFCs, but they have inspired DCFC designs. Figure 5.14 shows the "famous" cathode-supported tubular fuel

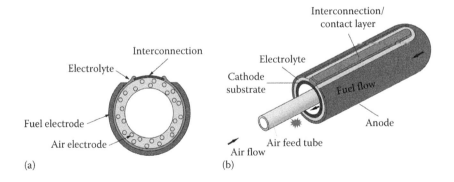

FIGURE 5.14 A schematic of single, cathode-supported tubular SOFC developed by Siemens–Westinghouse: (a) cross-sectional view; (b) side view of the fuel cell [126]. (Reprinted from *J. Power Sources*, 237, Huang, K. and Singhal, S.C., Cathode-supported tubular solid oxide fuel cell technology: A critical review, 84–97, Copyright 2013, with permission from Elsevier.)

cell with an oxidizing atmosphere sealed in the inner tube developed by Siemens–Westinghouse. This type of fuel cell demonstrates satisfactory capabilities of utilizing reformed natural gases as well as feasibilities in stacking [125]. One significant advantage of this cell configuration is that it collects current from cathodes under reducing atmospheres, and highly conductive metals can be used as both the anode and cathode current collectors.

The stacking of DCFCs can also be simplified if the anode is placed outward and the large anode chamber is used to house the reactor. Losses during current collection can be largely avoided if cathode current collection is performed in the anode chamber [127,128].

5.7.1.2   Molten Media Stability

Molten media such as molten carbonates are known to be highly corrosive. For example, Ni and NiO as common electrode materials were found to be unstable and slowly dissolved in these molten carbonates [129,130]. This etching effect can erode the cathodes of MC-DCFCs and Ni-based anodes of hybrid DCFCs. The compatibility between molten carbonates and solid-state materials in hybrid DCFCs is another challenge. Although YSZ electrolytes were thought to be stable in Li-K molten carbonates under reducing atmospheres, in reality, YSZ electrolytes were found to be etched under both inert and oxidizing atmospheres [11]. The oxidation of carbon and the transportation of oxidizing species during fuel cell operations can lead to a rise in oxygen partial pressures at certain local regions, making it a weak point for molten carbonates to attack the electrolyte. Alternative electrolyte materials such as GDC or less corrosive and conductive K-Na carbonate melts can be adopted to achieve better fuel cell stabilities [8].

The corrosion of electrolytes was also found in DCFCs using liquid metals as anodes. Pure Sb and Sb-Bi alloys were found to be corrosive in ScSZ electrolytes, while no significant thinning was found on YSZ electrolytes when liquid Sb was used as the anode. This higher stability of YSZ can be attributed to the larger ion radius of Y^{3+} as compared to Sc^{3+}, making it harder for doping ions to migrate from the electrolyte lattice to the liquid metal [98]. A later study on a similar test configuration further clarified this; YSZ electrolytes were observed being etched by liquid Sb anodes, and the corrosion effects were found to be closely related to the flow direction of the liquid metal [98,131]. Therefore, tubular fuel cells inserted into liquid Sb baths can be a promising method to mitigate the erosion of electrolytes when using liquid Sb as the electrolyte. In addition to fuel cell compartments, molten media also have negative effects on current collectors, such as metal mesh etching in molten carbonates and dissolving steel-based alloys in liquid metal baths. For example, metals such as bismuth (Bi) are highly corrosive, causing significant thinning of Re wires during testing of liquid Bi anode DCFCs.

5.7.1.3   Ash Content and Mineral Elements in Coal

Ash content in coal is the most challenging aspect of coal conversion in fuel cells. In current DCFC studies, most researchers use carbon black or de-ash coal as fuels to investigate carbon conversion kinetics as a means of simplifying the reaction system. However, ash-free carbon fuels (ash content around 0.2%) are 1.8 times more expensive than raw coal, making the handling of ash content an inevitable issue for direct carbon conversion. The accumulation of ash in both solid state and molten anodes ceases fuel cell reactions. Alkaline (IA group), alkaline earth (IIA) metals, and Fe in ash content are often regarded as promoters due to their catalytic roles in carbon gasification, whereas other species, such as alumina and silica, are considered inhibitors of DCFCs due to their roles in occupying reactive interfaces or blocking mass transports in porous anodes. These effects can be determined by adding respective compounds individually into DCFC anodes [78]. Their presence in coal along with the coeffects of several coexisting species and their interactions with the anode, electrolyte, and other fuel cell components are often left out of consideration.

A destructive experiment was carried out by Tao et al. [132] to study the poisoning effects of mineral elements on liquid Sn anode DCFCs; 400 ppm elemental As, Cr, V, and 200 ppm metallic Nb and Mo, totaling 1600 ppm of contaminates considered soluble in liquid Sn were added to a liquid Sn bath. The amount of the five elements added was higher than the worst-case scenario that a DCFC can meet when fed with real coal because these elements in coal are more likely to form chemical bonds rather than be in elemental forms. These elements in their oxides, silicates, sulfates, and chloride are much less soluble in liquid Sn than their elemental forms. Therefore, the experiment was carried out to demonstrate the accelerated poisoning effects of soluble contaminant elements on DCFCs. The performance of the DCFC with the contaminated liquid Sn anode indeed showed fast potential degradation rates of 3% per 100 hours due to anode contamination. Postmortem inspection of the YSZ electrolyte showed that Cr oxide as well as V oxide covered substantial areas of the electrolyte surface, with surface erosion and crystalline grain peeling-off also observed because of the contamination. Other ash contents such as oxides and sulfates of Si, Al, and Fe in the form of particulates or slag in the liquid Sn bath can be removed by filtration and skimming.

Aside from solid-state ash content, gaseous impurities can also be a threat to DCFC anodes. Elemental sulfur is the most intensively studied impurity due to its wide existence in both solid carbon fuels and gaseous fuels. Sulfur in coal exists in both organic and inorganic forms and binds to compounds migrating from the solid phase to the gas phase in the form of H_2S and COS under reducing atmospheres when the carbon fuel experiences gasification or pyrolysis. Thermodynamic predictions of Ni-based anodes poisoned by sulfur as a function of working potential and operation temperature were conducted based on a local equilibrium reaction model in an Ni–S–O–H–C system [133]. Effects of sulfur poisoning grow significantly if SOFC overpotentials increase due to the accumulation of sulfur content at the TPB due to the electrochemical oxidation of H_2S and the formation of Ni-S eutectic liquids. The formation of Ni–S compounds or even liquid-phase products will accelerate Ni migration, resulting in losses of catalysts, percolations, and electronic conductions of the anode [134]. In this aspect, sulfur-resistant anode materials, as well as their mechanisms, have been developed and reviewed [135,136].

Other coal-derived impurities such as PH_3, AsH_3, and HCl have also been found to cause negative effects on fuel cell operations, and the detailed poisoning effects have been reviewed by Cayan et al. [137]; 0.1 ppm of AsH_3 can result in decreased fuel cell performances due to the formation of less conductive AsNi phases in the anode region [138]. Because it reacts equally with Ni particles in the anode, arsenic is considered to be a less fatal threat than sulfur, mainly attacking the TPB region. Marina et al. [139] found that a PH_3 level as low as 2 ppm can lead to irreversible SOFC degradations at 800°C. Synergistic effects of different impurities are very complicated. Trembly et al. [140] reported a reversible decrease of 17.4% in fuel cell performances in a 100 h test at 800°C when 20 ppm of HCl was introduced into the anode chamber. However, Cl turned out to be a performance stabilizer if other contaminants were also present in the anode chamber. Adding Cl (in form of CH_3Cl) to a PH_3- or AsH_3-containing atmosphere will lower the rate of fuel cell degradation, as Cl can react with Ni–P alloys to form gaseous NiCl, resulting in the restructuring of catalyst surfaces [141]. If S is in the As/P system, the synergistic effects can be very destructive [142].

Syngas purification systems can also cause secondary pollution to fuel gases because warm gas cleanup processes are a more energy-efficient method to purify raw syngases. A ZnO reactor can be used as a terminal defense to collect S-containing species. Zn vapor can also be found in purified gases, and it is suspected to cause fuel cell degradations [137].

5.7.2 CONCLUSIONS

Solid oxide direct carbon fuel cells are power devices that can directly convert solid carbon fuels into power cleanly and efficiently. SO-DCFCs employ solid oxide ceramic as the electrolyte to conduct electrochemical conversions of carbon. However, carbon fuels are difficult to transport through the porous anodes of traditional fuel cells powered by gaseous or liquid fuels. Therefore, a lack of gas-phase reactants and separation between the carbon fuels and the electrochemically reactive sites result in limited cell performances of direct carbon fuel cells. To solve this problem, various methods have been developed to increase fuel cell power densities (1) using MIEC anodes to supply more O^{2-} to carbon; (2) using molten mediums such as metals and carbonates to improve contact conditions between the carbon and the anode, as well as to accelerate carbon fuel oxidation rates; and (3) using CO_2 or H_2O as gasification agents and catalysts in carbon beds to promote carbon gasification processes. Research has shown that the transport of oxygen to carbon surfaces is crucial for both gasification and electrochemical oxidation of carbon.

Major issues for current SO-DCFC technologies are the collection of current, the stability of molten media, and the poisoning effect of contaminants in carbon fuels. Solutions to resolve all of these challenges are still under development, with a focus on enhancing power densities, improving long-term stabilities and at the same time maintaining high efficiencies and low emissions.

REFERENCES

1. Li H., Liu Q., Li Y. 2010. A carbon in molten carbonate anode model for a direct carbon fuel cell. *Electrochimica Acta* 55(6): 1958–1965.
2. Gür T.M., 2010. Mechanistic modes for solid carbon onversion in high temperature fuel cells. *Journal of the Electrochemical Society* 157(5): B751–B759.
3. Shi Y. et al. 2007. Modeling of an anode-supported Ni–YSZ|Ni–ScSZ|ScSZ|LSM–ScSZ multiple layers SOFC cell Part I. Experiments, model development and validation. *Journal of Power Sources* 172(1): 235–245.
4. Zecevic S., Patton E.M., Parhami P. 2014. Carbon–air fuel cell without a reforming process. *Carbon* 42(10): 1983–1993.
5. Hackett G.A., Zondlo J.W., Svensson R. 2007. Evaluation of carbon materials for use in a direct carbon fuel cell. *Journal of Power Sources* 168(1): 111–118.
6. Cherepy N.J. et al. 2005. Direct conversion of carbon fuels in a molten carbonate fuel cell. *Journal of the Electrochemical Society* 152(1): A80–A87.
7. Cooper J.F. 2004. Direct Conversion of Coal and Coal-Derived Carbon in Fuel Cells. In *The 2nd International Conference on Fuel Cell Science, Engineering and Technology*. Rochester, NY.
8. Pointon K. et al. 2006. The development of a carbon–air semi fuel cell. *Journal of Power Sources* 162(2): 750–756.
9. Nabae Y., Pointon K.D., Irvine J.T.S. 2008. Electrochemical oxidation of solid carbon in hybrid DCFC with solid oxide and molten carbonate binary electrolyte. *Energy & Environmental Science* 1(1): 148.
10. Kaklidis N. et al. 2016. Effect of fuel thermal pretreament on the electrochemical performance of a direct lignite coal fuel cell. *Solid State Ionics* 288: 140–146.
11. Jiang C. et al. 2012. Demonstration of high power, direct conversion of waste-derived carbon in a hybrid direct carbon fuel cell. *Energy & Environmental Science* 5: 6973–6980.
12. Jain S.L. et al., 2008. Solid state electrochemistry of direct carbon/air fuel cells. *Solid State Ionics* 179(27–32): 1417–1421.
13. Jiang C., Irvine J.T.S. 2011. Catalysis and oxidation of carbon in a hybrid direct carbon fuel cell. *Journal of Power Sources* 196(17): 7318–7322.
14. Deleebeeck L., Hansen K.K. 2015. Hybrid direct carbon fuel cell performance with anode current collector material. *Journal of Electrochemical Energy Conversion and Storage* 12(6): 064501.
15. Xu X. et al. 2013. Optimization of a direct carbon fuel cell for operation below 700°C. *International Journal of Hydrogen Energy* 38(13): 5367–5374.
16. Myung J. et al. 2015. Nano-composite structural Ni–Sn alloy anodes for high performance and durability of direct methane-fueled SOFCs. *Journal of Material Chemistry A* 3(26): 13801–13806.
17. Yang Q. et al. 2016. Direct operation of methane fueled solid oxide fuel cells with Ni cermet anode via Sn modification. *International Journal of Hydrogen Energy* 41(26): 11391–11398.
18. Ju H. et al. 2012. Enhanced anode interface for electrochemical oxidation of solid fuel in direct carbon fuel cells: The role of liquid Sn in mixed state. *Journal of Power Sources* 198: 36–41.
19. Tao T. et al. 2007. Anode polarization in liquid tin anode solid oxide fuel cell. *ECS Transactions* 7(1): 1389–1397.
20. Jayakumar A., Vohs J.M., Gorte R.J. 2010. Molten-metal electrodes for solid oxide fuel cells. *Industrial & Engineering Chemistry Research* 49(21): 10237–10241.
21. Jayakumar A. et al. 2011. A direct carbon fuel cell with a molten antimony anode. *Energy & Environmental Science* 4(10): 4133–4137.

22. Toleuova A. et al. 2015. Mechanistic studies of liquid metal anode SOFCs I. Oxidation of hydrogen in chemical—Electrochemical mode. *Journal of the Electrochemical Society* 162(9): F988–F999.

23. Xu K., Li Z., Shi M. et al. 2017. Investigation of the anode reactions in SO-DCFCs fueled by Sn–C mixture fuels[J]. *Proceedings of the Combustion Institute*, vol. 36(3), 31 July–5 August 2016, Seoul, Korea, pp. 4435–4442.

24. Deleebeeck L., Hansen K.K. 2014. Hybrid direct carbon fuel cells and their reaction mechanisms—A review. *Journal of Solid State Electrochemistry* 18(4): 861–882.

25. Javadekar A. et al. 2012. Molten silver as a direct carbon fuel cell anode. *Journal of Power Sources* 214: 239–243.

26. Kulkarni A. et al. 2012. Mixed ionic electronic conducting perovskite anode for direct carbon fuel cells. *International Journal of Hydrogen Energy* 37(24): 19092–19102.

27. Li C., Shi Y., Cai N. 2011. Effect of contact type between anode and carbonaceous fuels on direct carbon fuel cell reaction characteristics. *Journal of Power Sources*, 196(10): 4588–4593.

28. Gür, T.M., Huggins R.A. 1992. Direct electrochemical conversion of carbon to electrical eEnergy in a high temperature fuel cell. *Journal of the Electrochemical Society* 139(10): L95–L97.

29. Mitchell R.E., Ma L., Kim B. 2007. On the burning behavior of pulverized coal chars. *Combustion and Flame* 151(3): 426–436.

30. Koenig P.C., Squires R.G., Laurendeau N.M. 1986. Char gasification by carbon dioxide: Further evidence for a two-site model. *Fuel* 65(3): 412–416.

31. Lee A.C., Mitchell R.E., Gür T.M. 2009. Modeling of CO_2 gasification of carbon for integration with solid oxide fuel cells. *AIChE Journal* 55(4): 983–992.

32. Mühlen H., van Heek K.H., Jüntgen H. 1985. Kinetic studies of steam gasification of char in the presence of H_2, CO_2 and CO. *Fuel* 64(7): 944–949.

33. Harris D.J., Smith I.W. 1991. Intrinsic reactivity of petroleum coke and brown coal char to carbon dioxide, steam and oxygen. *Symposium (International) on Combustion* 23(1): 1185–1190.

34. Zhou J. et al. 2012. A promising direct carbon fuel cell based on the cathode-supported tubular solid oxide fuel cell technology. *Electrochimica Acta* 74: 267–270.

35. Deng B. et al. 2014. Effect of catalyst on the performance of a solid oxide-based carbon fuel cell with an internal reforming process. *Fuel Cells* 14(6): 991–998.

36. Williams M.C., Strakey J., Sudoval W. 2006. U.S. DOE fossil energy fuel cells program. *Journal of Power Sources* 159(2): 1241–1247.

37. Rady A.C. et al. 2012. Review of fuels for direct carbon fuel cells. *Energy & Fuels* 26(3): 1471–1488.

38. Rady A.C. et al. 2015. Direct carbon fuel cell operation on brown coal with a Ni-GDC-YSZ anode. *Electrochimica Acta* 178: 721–731.

39. Dudek M., Tomczyk P. 2011. Composite fuel for direct carbon fuel cell. *Catalysis Today* 176(1): 388–392.

40. Xiao J. et al. 2016. Characterization of symmetrical SrFe0.75Mo0.25O3–δ electrodes in direct carbon solid oxide fuel cells. *Journal of Alloys and Compounds* 688: 939–945.

41. Suzuki T., Inoue K., Watanabe Y. 1988. Temperature-programmed deposition and CO_2-pulsed gasification of sodium- or iron-loaded yallourn coal char. *Energy & Fuels* 2(5): 673–679.

42. Tanaka S. et al. 1995. CO_2 gasification of iron-loaded carbons: Activation of the iron catalyst with CO. *Energy & Fuels* 9(1): 45–52.

43. Li X. et al. 2004. Volatilisation and catalytic effects of alkali and alkaline earth metallic species during the pyrolysis and gasification of Victorian brown coal. Part VI. Further investigation into the effects of volatile-char interactions. *Fuel* 83(10): 1273–1279.

44. Yu X., Shi Y., Cai N. 2015. Electrochemical impedance characterization on catalytic carbon gasification reaction process. *Fuel* 143: 499–503.
45. Tang Y., Liu J. 2010. Effect of anode and Boudouard reaction catalysts on the performance of direct carbon solid oxide fuel cells. *International Journal of Hydrogen Energy* 35(20): 11188–11193.
46. Li C., Shi Y., Cai N. 2010. Performance improvement of direct carbon fuel cell by introducing catalytic gasification process. *Journal of Power Sources* 195(15). 4660–4666.
47. Yu X. et al. 2014. Using potassium catalytic gasification to improve the performance of solid oxide direct carbon fuel cells: Experimental characterization and elementary reaction modeling. *Journal of Power Sources* 252: 130–137.
48. Wu Y. et al. 2009. A new carbon fuel cell with high power output by integrating with in situ catalytic reverse Boudouard reaction. *Electrochemistry Communications* 11(6): 1265–1268.
49. Yang B. et al. 2015. A carbon-air battery for high power generation. *Angewandte Chemie International Edition* 54(12): 3722–3725.
50. Zhong Y. et al. 2016. Process investigation of a solid carbon-fueled solid oxide fuel cell integrated with a CO_2-permeating membrane and a sintering-resistant reverse boudouard reaction catalyst. *Energy & Fuels* 30(3): 1841–1848.
51. Bai Y. et al. 2011. Direct carbon solid oxide fuel cell—A potential high performance battery. *International Journal of Hydrogen Energy* 36(15): 9189–9194.
52. Lee A.C. et al. 2008. Conversion of solid carbonaceous fuels in a fluidized bed fuel cell. *Electrochemical and Solid-State Letters* 11(2): B20–B23.
53. Li S. et al. 2008. Direct carbon conversion in a helium fluidized bed fuel cell. *Solid State Ionics* 179(27–32): 1549–1552.
54. Ong K.M., Ghoniem A.F. 2016. Modeling of indirect carbon fuel cell systems with steam and dry gasification. *Journal of Power Sources* 313: 51–64.
55. Panopoulos K.D. et al. 2006. High temperature solid oxide fuel cell integrated with novel allothermal biomass gasification. *Journal of Power Sources* 159(1): 570–585.
56. Williams M.C., Strakey J.P., Surdoval W.A. 2005. The U.S. Department of Energy, Office of Fossil Energy Stationary Fuel Cell Program. *Journal of Power Sources* 143(1–2): 191–196.
57. Iyengar A.K.S., Newby R.A., Keairns D.L. November 24, 2014. *Techno-Economic Analysis of Integrated Gasification Fuel Cell Systems*. National Energy Technology Laboratory, USA. DOE/NETL-341/112613.
58. Newby R., Keairns D. February 22, 2011. *Analysis of Integrated Gasification Fuel Cell Plant Configurations*. National Energy Technology Laboratory, USA. DOE/NETL-2011-1482.
59. Pinkerton L., Varghese E., Woods M. August 2012. *Updated Costs (June 2011 Basis) for Selected Bituminous Baseline Cases*. National Energy Technology Laboratory, USA. DOE/NETL-341/082312.
60. Fout T. et al. July 6, 2015. *Cost and Performance Baseline for Fossil Energy Plants Volume 1a: Bituminous Coal (PC) and Natural Gas to Electricity Revision 3*. National Energy Technology Laboratory, USA. DOE/NETL-2015/1723. 2015.
61. Yoshikawa M. et al. 2006. Experimental determination of effective surface area and conductivities in the porous anode of molten carbonate fuel cell. *Journal of Power Sources* 158(1): 94–102.
62. Peelen W.H.A. et al. 2000. Electrochemical oxidation of carbon in a 62/38 mol% Li/K carbonate melt. *Journal of Applied Electrochemistry* 30(12): 1389–1395.
63. Li X. et al. 2010. Modification of coal as a fuel for the direct carbon fuel cell. *The Journal of Physical Chemistry A* 114(11): 3855–3862.
64. Li X. et al. 2009. Surface modification of carbon fuels for direct carbon fuel cells. *Journal of Power Sources* 186(1): 1–9.

65. Chen J.P., Wu S. 2004. Acid/base-treated activated carbons: Characterization of functional groups and metal adsorptive properties. *Langmuir* 20(6): 2233–2242.
66. Chen C.C. et al. 2012. Wetting behavior of carbon in molten carbonate. *Journal of the Electrochemical Society* 159(10): D597–D604.
67. Cooper J.F., Selman J.R. 2009. Electrochemical oxidation of carbon for electric power generation: A review. *ECS Transactions* 19(14): 15–25.
68. Chen M. et al. 2010. Carbon anode in direct carbon fuel cell. *International Journal of Hydrogen Energy* 35(7): 2732–2736.
69. Jain S.L. et al. 2009. Electrochemical performance of a hybrid direct carbon fuel cell powered by pyrolysed MDF. *Energy & Environmental Science* 2(6): 687–693.
70. Tulloch J. et al. 2014. Influence of selected coal contaminants on graphitic carbon electro-oxidation for application to the direct carbon fuel cell. *Journal of Power Sources* 260: 140–149.
71. Castellano M. et al. 2010. Bulk and surface properties of commercial kaolins. *Applied Clay Science* 48(3): 446–454.
72. Allen J.A., Glenn M., Donne S.W. 2015. The effect of coal type and pyrolysis temperature on the electrochemical activity of coal at a solid carbon anode in molten carbonate media. *Journal of Power Sources* 279: 384–393.
73. Kojima T. et al. 2008. Density, surface tension, and electrical conductivity of ternary molten carbonate system Li_2CO_3-Na_2CO_3-K_2CO_3 and methods for their estimation. *Journal of The Electrochemical Society* 155(7): F150–F156.
74. Kouchachvili L., Ikura M. 2011. Performance of direct carbon fuel cell. *International Journal of Hydrogen Energy* 36(16): 10263–10268.
75. Kojima T. et al. 1999. The surface tension and the density of molten binary alkali carbonate systems. *Electrochemistry* 67(6): 593–602.
76. Nabae Y., Pointon K.D., Irvine J.T.S. 2009. Ni/C slurries based on molten carbonates as a fuel for hybrid direct carbon fuel cells. *Journal of the Electrochemical Society* 156(6): B716–B720.
77. Cooper J.F., Selman J.R. 2012. Analysis of the carbon anode in direct carbon conversion fuel cells. *International Journal of Hydrogen Energy* 37(24): 19319–19328.
78. Li X. et al. 2010. Evaluation of raw coals as fuels for direct carbon fuel cells. *Journal of Power Sources* 195(13): 4051–4058.
79. Ahn S.Y. et al. 2013. Utilization of wood biomass char in a direct carbon fuel cell (DCFC) system. *Applied Energy* 105: 207–216.
80. Lee C.G., Song M.B. 2012. Carbon oxidation with electrically insulated carbon fuel in A coin type direct carbon fuel cell. *Fuel Cells* 12(6): 1042–1047.
81. Li X. et al. 2008. Factors that determine the performance of carbon fuels in the direct carbon fuel cell. *Industrial & Engineering Chemistry Research* 47(23): 9670–9677.
82. Gorte R.J., Vohs J.M. 2009. Nanostructured anodes for solid oxide fuel cells. *Current Opinion in Colloid & Interface Science* 14(4): 236–244.
83. Lipilin A.S. et al. 2012. Liquid anode eelectrochmical cell US Patent, US 8101310 B2.
84. SRI International. 2006. SRI unveils direct carbon fuel cell technology. *Fuel Cells Bulletin* 2006(1): 11.
85. Wang H., Shi Y., Cai N. 2013. Effects of interface roughness on a liquid-Sb-anode solid oxide fuel cell. *International Journal of Hydrogen Energy* 38(35): 15379–15387.
86. Javadekar A. et al. 2012. Energy storage in electrochemical cells with molten Sb electrodes. *Journal of the Electrochemical Society* 159(4): A386.
87. van Arkel A.E., Flood E.A., Bright N.F.H. 1953. The electrical conductivity of molten oxides. *Canadian Journal of Chemistry* 31(11): 1009–1019.
88. Cao T. et al. 2015. Numerical simulation and experimental characterization of the performance evolution of a liquid antimony anode fuel cell. *Journal of Power Sources* 284: 536–546.

89. Wang H., Shi Y., Cai N. 2014. Polarization characteristics of liquid antimony anode with smooth single-crystal solid oxide electrolyte. *Journal of Power Sources* 245: 164–170.

90. Xu X.Y., Zhou W., Zhu Z.H. 2013. Samaria-doped ceria electrolyte supported direct carbon fuel cell with molten antimony as the anode. *Industrial & Engineering Chemistry Research* 52(50): 17927–17933.

91. Aqra F., Ayyad A. 2011. Theoretical estimation of temperature-dependent surface tension of liquid antimony, boron, and sulfur. *Metallurgical and Materials Transactions B* 42(3): 437–440.

92. Cao T., Shi Y., Cai N. 2016. Liquid antimony anode fluidization within a tubular direct carbon fuel cell. *Journal of the Electrochemical Society* 163(3): F127–F131.

93. Davis J.K., Bartell F.E. 1948. Determination of suface tension of molten materials—Adaptation of the pendant drop method. *Analytical Chemistry* 20(12): 1182–1185.

94. Jayakumar A. et al. 2010. A comparison of molten Sn and Bi for solid oxide fuel cell anodes. *Journal of the Electrochemical Society* 157(3): B365–B369.

95. Javadekar A., Jayakumar A., Gorte R.J. et al. 2011. Characteristics of molten alloys as anodes in solid oxide fuel cells. *Journal of the Electrochemical Society*, 158(12): B1472–B1478.

96. Wang H., Shi Y., Cai N. 2014. Characteristics of liquid stannum anode fuel cell operated in battery mode and CO/H$_2$/carbon fuel mode. *Journal of Power Sources* 246: 204–212.

97. Otaegui L. et al. 2014. Performance and stability of a liquid anode high-temperature metal–air battery. *Journal of Power Sources* 247: 749–755.

98. Jayakumar A. et al. The stability of direct carbon fuel cells with molten Sb and Sb-Bi alloy anodes. *AIChE Journal* 59(9): 3342–3348.

99. Tao T. et al. 2007. Liquid tin anode solid oxide fuel cell for direct carbonaceous fuel conversion. *ECS Transactions* 5(1): 463–472.

100. McPhee W.A. et al. 2009. Demonstration of a liquid-tin anode solid-oxide fuel cell (LTA-SOFC) operating from biodiesel fuel. *Energy & Fuels* 23(10): 5036–5041.

101. Cao T. et al. 2014. Direct carbon fuel conversion in a liquid antimony anode solid oxide fuel cell. *Fuel* 135: 223–227.

102. Cao T., Shi Y., Cai N. 2015. Introducing anode fluidization into a tubular liquid antimony anode direct carbon fuel cell. *ECS Transactions* 68(1): 2703–2712.

103. Gorte R.J., Oh T. 2013. Direct carbon fuel cell and stack designs, US patent, US20160156056.

104. Tao T. et al. 2008. Advancement in liquid tin anode-solid oxide fuel cell technology. *ECS Transactions* 12(1): 681–690.

105. Carlson E.J. et al. December 2006. *Assessment of a Novel Direct Coal Conversion—Fuel Cell Technology for Electric Utility Markets.* Electric Power Research Institute, Palo Alto, CA.

106. Gopalan S., Ye G., Pal U.B. 2006. Regenerative, coal-based solid oxide fuel cell-electrolyzers. *Journal of Power Sources* 162(1): 74–80.

107. Cooper J.F., Krueger R. April 15–16, 2003. Direct carbon (coal) conversion batteries and fuel cells. In *Fourth Annual SECA Meeting.* Seattle, WA.

108. Chen T.P. April 2008. *Program on Technology Innovation: Systems Assessment of Direct Carbon Fuel Cells Technology.* Electric Power Research Institute, Palo Alto, CA.

109. Lee A.C., Mitchell R.E., Gür T.M. 2009. Thermodynamic analysis of gasification-driven direct carbon fuel cells. *Journal of Power Sources* 194(2): 774–785.

110. Armstrong G.J. et al. 2013. Modeling heat transfer effects in a solid oxide carbon fuel cell. *ECS Transactions* 50(45): 143–150.

111. Tao T. July 27–29, 2010. Novel fuel cells for coal based systems. In *11th Annual SECA Workshop.* Pittsburg, PA.

112. Tao T. July 25–28, 2011. Direct coal conversion in liquid tin anode SOFC. In *12th Annual SECA Workshop.* Pittsburg, PA.

113. Campanari S., Gazzani M., Romano M.C. 2013. Analysis of direct carbon fuel cell based coal fired power cycles with CO_2 capture. *Journal of Engineering for Gas Turbines and Power* 2013(1): 011701.

114. Ghezel-Ayagh H. July 14–16, 2009. Solid oxide fuel cell program at fuel cell energy Inc. In *10th Annual SECA Workshop*. Pittsburgh, PA.

115. Tam S.S. July 27–29, 2010. Clean Coal Energy Research. In *11th Annual SECA Workshop*. Pittsburg, PA.

116. Greppi P., Bosio B., Arato E. 2009. Feasibility of the integration of a molten carbonate fuel-cell system and an integrated gasification combined cycle. *International Journal of Hydrogen Energy* 34(20): 8664–8669.

117. Bang-Møller C., Rokni M., Elmegaard B. 2011. Exergy analysis and optimization of a biomass gasification, solid oxide fuel cell and micro gas turbine hybrid system. *Energy*, 36(8): 4740–4752.

118. Ghezel-Ayagh H. et al. November 16–19, 2009. Integrated coal gasification and solid oxide fuel cell systems for centralized power generation. In *Fuel Cell Seminar & Exposition*. Palm Springs, CA.

119. Ghezel-Ayagh H. 2010. Integrated coal gasification solid oxide fuel cell systems. Presented to *Hydrogen and Fuel Cell Technical Advisory Committee*. Washington, DC.

120. Franklin H.D. et al. November 14, 1982. Dynamic Simulation of Exxon's Catalytic Coal-Gasification Process. *Annual Meeting of Heat Transfer and Energy Conversion*. Los Angeles, CA.

121. Iyengar A. et al. July 22, 2014. *IGFC and NGFC Pathway Studies-Estimation of Stack Degradation Costs and Salient Results*. National Energy Technology Laboratory, USA. DOE/NETL-2016/1766.

122. Sikarwar V.S. et al. 2016. An overview of advances in biomass gasification. *Energy & Environmental Science*, 9(10): 2939–2977.

123. Zhang H. et al. 2014. Performance analysis of a direct carbon fuel cell with molten carbonate electrolyte. *Energy* 68: 292–300.

124. Liu Q. et al. 2008. Modeling and simulation of a single direct carbon fuel cell. *Journal of Power Sources* 185(2): 1022–1029.

125. George R.A. 2000. Status of tubular SOFC field unit demonstrations. *Journal of Power Sources* 86(1–2): 134–139.

126. Huang K., Singhal S.C. 2013. Cathode-supported tubular solid oxide fuel cell technology: A critical review. *Journal of Power Sources* 237: 84–97.

127. Yoshida, Y., Hisatome N., Takenobu K. 2003. Development of SOFC for products. Mitsubishi Heavy Industries, Ltd. *Technical Review* 40(4): 1–5.

128. Vora S.D. July 15, 2009. SECA Program Review of Siemens Energy. In *10th Annual SECA Workshop*. Pittsburgh, PA.

129. Cassir M. et al. 1998. Thermodynamic and electrochemical behavior of nickel in molten Li_2CO_3—Na_2CO_3 modified by addition of calcium carbonate. *Journal of Electroanalytical Chemistry* 452(1): 127–137.

130. Kudo T. et al. X-ray diffractometric study of in situ oxidation of Ni in Li/K and Li/Na carbonate eutectic. *Journal of Power Sources* 104(2): 272–280.

131. Zhou X. et al. 2015. Zirconia-based electrolyte stability in direct-carbon fuel cells with molten Sb anodes. *Journal of the Electrochemical Society* 162(6): F567–F570.

132. Tao T. et al. 2009. Liquid tin anode SOFC for direct fuel conversion impact of coal and JP-8 impurities. *ECS Transactions* 25(2): 1115–1124.

133. Kishimoto H. et al. 2010. Sulfur poisoning on SOFC Ni anodes: Thermodynamic analyses within local equilibrium anode reaction model. *Journal of The Electrochemical Society* 157(6): B802–B813.

134. Lussiera A. et al. 2008. Mechanism for SOFC anode degradation from hydrogen sulfide exposure. *International Journal of Hydrogen Energy* 33(14): 3945–3951.

135. Cheng Z., Zha S., Liu M. 2006. Stability of materials as candidates for sulfur-resistant anodes of solid oxide fuel cells. *Journal of the Electrochemical Society* 153(7): A1302.

136. Gong M. et al. 2007. Sulfur-tolerant anode materials for solid oxide fuel cell application. *Journal of Power Sources* 168(2): 289–298.

137. Cayan F.N. et al. 2008. Effects of coal syngas impurities on anodes of solid oxide fuel cells. *Journal of Power Sources* 185(2): 595–602.

138. Trembly J.P., Gemmen R.S., Bayless D.J. 2007. The effect of coal syngas containing AsH3 on the performance of SOFCs: Investigations into the effect of operational temperature, current density and AsH3 concentration. *Journal of Power Sources* 171(2): 818–825.

139. Marina O.A. et al. 2010. Degradation mechanisms of SOFC anodes in coal gas containing phosphorus. *Solid State Ionics* 181(8–10): 430–440.

140. Trembly J.P., Gemmen R.S., Bayless D.J. 2007. The effect of coal syngas containing HCl on the performance of solid oxide fuel cells: Investigations into the effect of operational temperature and HCl concentration. *Journal of Power Sources* 169(2): 347–354.

141. Bao J. et al. 2010. Impedance study of the synergistic effects of coal contaminants: Is Cl a contaminant or a performance stabilizer. *Journal of the Electrochemical Society* 157(3): B415–B424.

142. Bao J. et al. 2010. Effect of various coal gas contaminants on the performance of solid oxide fuel cells: Part III. Synergistic effects. *Journal of Power Sources* 195(5): 1316–1324.

Index

Printed and bound by CPI Group (UK) Ltd, Croydon, CR0 4YY

01/11/2024

01782619-0001